Keep it Running, Keep it Safe

Keep it Running, Keep it Safe

by

William Wong
CEng, FIMechE, FIMarEST

**Professional
Engineering
Publishing**

Professional Engineering Publishing Limited
London and Bury St Edmunds, UK

First published 2004

© W Wong

ISBN 1 86058 407 1

A CIP catalogue record for this book is available from the British Library.

Front cover illustration courtesy of Nuovo Pignone.

Printed and bound in Great Britain by The Cromwell Press Limited, Wiltshire, UK

Dedication

To my old friends and colleagues, fellow travellers,
the Rotating Machinery Engineers.
Especially to Jan Smulders who set me on the road.
And to those that follow, may they benefit from our lessons learnt.

Contents

Acknowledgements   *xiii*

How to Use the Guide   *xv*

Chapter 1 Scope and Introduction   **1**
  1.1 Scope   1
  1.2 Introduction   1
  1.3 The evolution of health and safety legislation   2
      1.3.1 Evolution of the law to include safety in design   4
      1.3.2 The duty of compliance   5
      1.3.3 Enforcement of the law   6
  1.4 Codes and standards   7

Chapter 2 The Law   **9**
  2.1 Introduction   9
  2.2 Construction (Design and Management) (CDM) Regulations   10
      2.2.1 The client (on initiation of a project)   11
      2.2.2 The planning supervisor   11
      2.2.3 The designer   12
      2.2.4 The client (as operator)   12
  2.3 Dangerous Substances and Explosive Atmospheres Regulations (DSEAR) 2002   13
      2.3.1 Hazardous area classification   13
      2.3.2 Risk management   16
      2.3.3 Required documentation   17
  2.4 The Equipment and Protective Systems Intended for Use in Potentially Explosive Atmospheres Regulations 1996 (S. I. 1996/192), (ATEX Directive 94/9/EC, as amended 2001)   18

2.4.1	Introduction	18
2.4.2	The Essential Health and Safety Requirements (EHSRs)	18
2.4.3	Equipment groups and categories	19
2.4.4	Conformity assessment requirements	20
2.4.5	Mechanical equipment	21

2.5   Provision and Use of Work Equipment Regulations (PUWER) 1992   22

2.6   Machinery Directive {[Supply of Machinery (Safety) Regulations] 1992, as amended 1994}   23

2.6.1	Introduction	23
2.6.2	Individual machines and safety components	24
2.6.3	Assemblies of machines	25
2.6.4	Machinery for incorporation	26
2.6.5	Technical file	27
2.6.6	Confidentiality	30
2.6.7	Machinery modifications	30
2.6.8	Documentation	31
2.6.9	Summary of actions required by the regulations and amendments	31
2.6.10	Machinery for process plant	32

2.7   Electromagnetic Compatibility (EMC) Regulations 1992   33

2.8   The Electrical Equipment (Safety) Regulations 1994   34

2.9   Pressure Equipment Directive (PED)   36

2.9.1	Introduction	36
2.9.2	Harmonized standards	37
2.9.3	The EHSRs	38
2.9.4	Conformity assessment procedure	39
2.9.5	Compliance with the EHSRs	46

2.10   Simple Pressure Vessels Directive (SPVD)   48

2.11   Pressure Systems Safety Regulations 2000 (PSSR)   49

2.12   Control of Major Accident Hazards (COMAH) Regulations 1999   49

2.13   Summary   50

2.13.1 Example of a gas turbine enclosure   50

Chapter 3   Risk Assessment   53

3.1   Introduction   53

3.2   The information required   54

3.3   Identification of hazards   54

	3.3.1	Machine hazard assessment	55
	3.3.2	Maintenance hazards	55
	3.3.3	Human interface hazards	56
	3.3.4	Process materials hazard assessment	57
	3.3.5	Transient and operating location hazards	57
3.4	Risk assessment		58
3.5	Acceptable risk – ALARP		61
3.6	Failure mode and effects analysis (FMEA)		61
3.7	Hazard and operability (HAZOP) studies		62
3.8	Summary		63

Chapter 4	**Reliability Assessment**		**65**
4.1	Introduction		65
4.2	Factors that affect the probability of failure		66
4.3	Measures to reduce the risk of failure		66
4.4	Reliability and failure probability		67
	4.4.1	Failure rate (failure frequency per unit time)	67
	4.4.2	Environmental and stress factors	68
	4.4.3	Probability of failure	68
	4.4.4	Reliability prediction	69
	4.4.5	Operational reliability	70
	4.4.6	Installed spare units (redundancy)	70
	4.4.7	Probability of failure on demand	71
	4.4.8	Common mode failure	71
	4.4.9	Fractional redundancy	72
	4.4.10	Control system reliability	72
4.5	Fault-tree analysis (FTA)		73
4.6	Factors affecting reliability assessment		74
4.7	Reliability in design		74
4.8	Reliability specification		78
4.9	Model safety and reliability specification		79
4.10	Summary		80
4.11	Reliability analysis – an illustration		81
	4.11.1	The scenario	81
4.12	References		86

Chapter 5	**Control of Risks**		**87**
5.1	Introduction		87
5.2	Responsibilities		87
	5.2.1	The client	87
	5.2.2	The designer	87

5.2.3   Project responsibilities   88
5.2.4   Safety planning supervisor   88
5.2.5   Machine design responsibilities   89
5.2.6   Safety or loss-prevention engineer   89
5.2.7   The machine manufacturer   89
5.2.8   The construction contractor   90
5.2.9   The construction/commissioning engineer   90
5.2.10 The site (plant) manager   90
5.3   Classification of machines   90
5.3.1   Significance of classification   92
5.4   Machine verification   92
5.4.1   Machine purpose and specification   93
5.4.2   Machine operation and protection   94
5.4.3   Machine maintenance and operation   95
5.4.4   Hazard review   97
5.4.5   Containment review   97

Chapter 6 Operations and Maintenance   99
6.1   Introduction   99
6.2   Handover   99
6.3   Reliability-centred maintenance (RCM)   100
6.4   Operations and maintenance   101
6.5   Inspection   103
6.6   System audits   103
6.7   Management of change   104
6.7.1   Modifications   104
6.7.2   The assessment of change   104
6.7.3   Compliance with requirements   104
6.8   Approval of modifications   105
6.8.1   Checks and reviews   105
6.8.2   The EH&S assessment form   105
6.8.3   Initialling and dating   106
6.8.4   Recording of details   106
6.9   Design standards   106

Chapter 7 Keep it Running   107
7.1   Introduction   107
7.1.1   Machine reliability   108
7.1.2   Dependability management   108
7.1.3   Process hazards   109
7.1.4   Manufacturing hazards   110

	7.2	Compression machinery	111
		7.2.1 Centrifugal compressors	112
		7.2.2 Axial compressors	118
		7.2.3 Screw compressors	119
		7.2.4 Reciprocating compressors	125
	7.3	Centrifugal pumps	137
	7.4	Drivers	144
		7.4.1 The machinery train	144
		7.4.2 Special-purpose steam turbines	145
		7.4.3 General-purpose steam turbines	153
		7.4.4 Gas turbines	154
		7.4.5 Synchronous motors	157
		7.4.6 Diesel engines	157
	7.5	Summary	158
	7.6	References	158

Chapter 8 Keep it Safe **161**

	8.1	Introduction	161
	8.2	Control and monitoring systems	161
		8.2.1 Instrument reliability	161
		8.2.2 Maintenance provisions	164
		8.2.3 Software development	164
	8.3	Pressure systems	166
		8.3.1 Introduction	166
		8.3.2 Over-pressure	167
		8.3.3 Materials failure	167
		8.3.4 Corrosion	170
		8.3.5 Corrosion protection	175
	8.4	Keeping it safe	175
		8.4.1 Introduction	175
		8.4.2 Failure statistics	176
		8.4.3 Risk assessment	177
		8.4.4 Risk-based inspection (RBI)	178
		8.4.5 A written scheme of examination	180
		8.4.6 Inspection requirements	181
		8.4.7 The management of RBI	189
	8.5	A warning note	190
	8.6	Summary	191
	8.7	References	191

Chapter 9 Epilogue   **193**
    9.1   Introduction   193
    9.2   An air compressor failure   193
        9.2.1   Gear failure   193
    9.3   The failure of a liquid oxygen pump to pump   194
        9.3.1   Failure to pump   195
    9.4   A problem of the twenty-first century   195
        9.4.1   Two-shift operations   196
        9.4.2   The investigation   196
        9.4.3   Investigation results   198
    9.5   Summary   199
    9.6   References   202

Bibliography   **203**

Directory   **207**

Appendix A Glossary   **217**

**Appendix B Materials safety data sheet – some common
                hazards**   **221**

**Appendix C Machinery Directive – checklist for process
                machinery**   **229**

Appendix D Documentation relating to a fire-water pump   **239**

Index   **277**

Acknowledgements

Second edition

The year 2002 was a watershed in the design and construction of process plant. It was the year when full compliance with the Pressure Equipment Regulations was required. As a result the contents of this book became out of date and an extensive revision was undertaken for this new edition. This second edition has benefited from the many suggestions and comments made by friends and old colleagues who aided the task of proof reading and making it simple. My thanks are especially due to:

- Robert Flood for proof reading.
- Mrs S. L. Cave-Jones and Dirk Wong for ideas on restructuring Chapters 3 and 4. Trusting that it will help the reader to swallow all the mathematics.
- Frank Ibbott, Klaus Lelgemann of GHH, Nick Muntz of Weir Pumps, Frank Rhodes, and Robin Wilson of CPI for their contributions to Chapter 7.
- Peter Kemp and Toby Lant of ERA Technology, for their contributions to Chapters 8 and 9.
- Dr Jian-Zhong Zhang for his help with the metallurgy in Chapter 8.
- GE Power Systems, Nuovo Pignone, for the cover photo.
- Siemens Power Generation, for the photos of steam turbine blade deposits.
- Compressor Products International (CPI) for the photos of compressor valves.
- Anonymous for the photos of pump vane cavitation damage.

First edition

The idea for this guide was conceived during the course of a one-day seminar, Maximizing Rotating Reliability, organized by the Mechanical Reliability Committee of the Process Industries Division of the IMechE, held on the 8th December 1994 at Headquarters. A working party was thereby set up with the task of planning the layout of the guide and providing material for its contents. The working party was made up of the following members.

J. Harris	University of Manchester
K. G. Rayner	ICI Engineering Technology
J. Lewis	ICI Eutech
D. T. Parr	AEA Technology
E. Parry	AEA Technology
J. Worsley	AEA Technology
W. Wong	Bechtel Ltd
D. Holland	Bechtel Ltd

Thanks are due to all the companies of the individuals concerned for the free use of their facilities during the course of the work.

Special thanks are due to R. Moss for valuable comments and to D. Heckle for proof reading.

The final proof copy was sent to a number of manufacturers.

The comments received from Weir Pumps were invaluable, especially their plea to reduce the Fog Index. Where some of the legal English is difficult, examples of interpretation have been added for clarification. We are also indebted to them for providing the compliance checklist (updated and revised by the author for the second edition) and the example of its application on a fire-water pump as supplied for a typical offshore project. Some useful comments were received from the British Pump Manufacturers Association and use has been made of material from the British Chemical Engineering Contractors Association and the Engineering Equipment and Material Users Association. It was also confirmed by the electric motor manufacturers that their interpretation is that all motors will be supplied as incorporated machines (second edition note: standard off-the-shelf motors are CE marked).

Last but not least thanks are also due to J. Harris, the Chairman of the Mechanical Reliability Committee, for all his work and encouragement.

How to Use the Guide

The guide is intended for the process industries. It provides a general interpretation of the Machinery Directive and other related legislation. The underlying objective is to ensure a common approach to health and safety. The guide meets this by outlining a procedure for the assessment of machine safety and reliability.

Public concern on health and safety issues has resulted in political action. There is now an ever-increasing enactment of safety, health, and environmental legislation throughout the world. Experience has shown that the previously prescriptive regulations are inadequate. This has resulted in new legislation that is objective and therefore open ended.

In the past, adhering to a checklist of requirements could satisfy the requirements for health and safety. The need for objective safety requires two undertakings to be demonstrated. First, that there has been sufficient effort to identify all risks to health and safety. Second, that all reasonable and practical measures have been taken for their reduction. This is in addition to meeting all the prescribed regulations such as the Health and Safety at Work Act in the UK and similar regulations elsewhere. These relate to fire precautions, safeguarding of moving parts, dangerous substances, and the like.

In the section on legal requirements, an overview is provided of the UK regulations that reflect the European Community Directives relevant to process plant. Some workable principles of interpretation are suggested. Interpretation and legislation enacting the European Commission Directives will differ in each member state. The regulations applicable in the country of interest must be consulted. All interpretations are only opinions, even those of government departments! Only the courts can make authoritative rulings on the correct interpretation, with the ultimate authority being the European Court of Justice.

The methods by which risk to health and safety can be identified and quantified are given in the sections on Hazard and Reliability Assessment. Sufficient detail is given for the reader to become aware of the concepts and procedures commonly in use. The information needed for their implementation is identified. This will enable a data package to be prepared for assessment. Where necessary consultants can be used to carry out the assessment.

The risk to health and safety will vary depending on the type of fluid being handled and the consequences of failure. In the chapter on the Control of Hazards a classification procedure to take account of this is proposed. A verification procedure is also given as a means of providing a final audit that all risks have been recognized.

A reference list, a reading list, and a list of videos is given for further study.

The Appendix contains a list of definitions, suggested data sheets for materials safety and checking machinery health and safety compliance.

An index list from the technical file relating to a fire-water pump for an offshore platform is also provided with some sample documentation. These show the use of the compliance checklist, the application of failure mode and effects analysis (FMEA) for hazard identification, and actions taken as a result of a HAZOP review.

Second edition comments
The opportunity has been taken to re-edit and enhance the book. Some of the chapters have been rewritten and two new chapters have been added. In particular:

Chapter 2, The Law, has been brought up to date as of the year 2003. This was the first year of full implementation of the Pressure Equipment Directive. This together with the ATEX Directive has now been included. The relationship between the different directives and their compliance in the context of process plant design is outlined.

Chapter 3 has been renamed Risk Assessment in line with current usage in the EU. An introduction to qualitative risk assessment has been provided together with the principle of 'as low as reasonably practicable' (ALARP) as required by the Health and Safety Executive regulations.

Chapter 4, Reliability Assessment, has been extended in scope to provide some elementary methods of quantitative analysis with an illustration on how they should be applied. These methods explain the theoretical basis of many of the testing and inspection practices that are essential for reliable operation.

Chapter 7, Keep it Running, has been added to provide information on machinery selection, their respective failure modes and operating characteristics. The situations that cause failure are outlined together with the measures that are needed for their avoidance.

Chapter 8, Keep it Safe, is concerned with avoiding the risk of failure due to instrumentation and pressure systems.

It is hoped that the addition of these last two chapters will bring together the requirements of the law and the practical steps that are needed for safe and reliable operation.

Chapter 9 has been added by way of an epilogue that gives examples that illustrate the interdependence of a machine with its pressure system and the effects of any change in operating conditions.

The Bibliography has been extended to provide more useful sources of information and services.

Chapter 1

Scope and Introduction

1.1 Scope

To provide a guide to good practice for the assessment of machine system reliability and hazards so as to:

- enable reliability (as related to safety) to be improved;
- enable hazards to be eliminated or controlled in a cost-effective manner;
- satisfy legislation, relative to the enactment of European Commission Directives, by the provision of guidance on the appropriate records and documentation, and to show the interrelationship between the relevant directives;
- develop an adequate cost-effective maintenance strategy for process plant machinery and pressure systems.

1.2 Introduction

There is a legal requirement to control the hazards associated with machinery throughout the lifetime of the equipment.

The law now requires the design, operation, and maintenance of machinery to be subject to a special review. This review is to enable the requirements for safety to be identified and prescribed. Safe use of the equipment throughout its operating life is the desired objective.

The reliability of a system or component is measured by its expected failure rate. Reliability becomes important in situations where the associated failure modes result in danger to health and safety. Another aspect of reliability is its impact on lost production. Reliable operation is of equal importance both to health and safety, and profitable operation.

In the process industries the loss of containment of fluids is of major concern. This is because in most cases it will pose a threat to health, safety, and/or the environment. Users of machinery in those industries will find this guide particularly useful.

There are established codes and standards applicable to the supply and construction of the machinery and plant for the power and process industries. There are also established practices in safety assessment and identification that are commonly used in the process industry. These have, in general, provided safe and reliable plant and machinery. But accidents still occur. It is believed that these accidents are mostly due to failure to understand application limits and maintenance requirements.

This gap in communication has been recognized in current (2003) and recent UK legislation. Both the Machinery Directive and the Construction (Design and Management) (CDM) Regulations are good examples of this. Both of these are concerned with the adequate management and systematic recording of provisions for safe and reliable plant operation. These together with the Pressure Equipment Regulations and ATEX cover all aspects of plant design and construction. The revised Pressure Systems Safety Regulations and the Control of Major Accident Hazards Regulations cover plant operations.

This guide identifies the practical measures required for compliance with the regulations, i.e. those measures that will provide adequate assessment of plant machinery safety and reliability. Users are warned, however, that the regulations are subject to interpretation, which, until tested in court, provides no clear code of practice.

Hazardous situations caused by incorrect interaction between static plant and machinery need to be identified and avoided as far as possible. The CDM Regulations place a duty on the plant designer to ensure that the interfaces between the plant and its machines are kept safe. All those conditions of operation that can reasonably be foreseen must be taken into account; guidance on these conditions is given in this latest edition of the book.

As an introduction to the law it may be of interest to know how present-day law has evolved.

1.3   The evolution of health and safety legislation

The onset of the industrial age resulted in more and more people working with machines and industrial processes. Work-related problems of health and safety became a serious concern. During the twentieth century it was thought that the fault was with the factory owner. The laws enacted were based on the need for the factory owner to provide a safe and healthy working environment. Laws were passed based on specific industry problems as they appeared. Things came to a head in the 1960s and the situation was then tackled by the enactment of an all-embracing law to protect the health and safety of workers.

In the USA, the Occupational Safety and Health Act (OSHA) was enacted in 1970, followed by the Process Safety Management of Highly Hazardous Chemicals, OSHA standard, Federal Regulation 1910–119 in 1992. This regulation was aimed at the employer who was required to:

- provide a file on process safety information;
- carry out a process hazard analysis;
- produce written operating procedures;
- ensure the training of operators;
- provide a pre-start-up safety review;
- ensure the integrity of mechanical equipment and have available supporting documentary evidence to show this;
- install a hot-work permit system;
- manage any change to ensure compliance with all the requirements as listed above;
- investigate accidents to report and take action to prevent or avoid them;
- carry out emergency planning and response procedures;
- conduct a Compliance Audit every three years.

In the UK, the Health and Safety at Work Act 1974 also centred on the employer with the duty to provide a safe and healthy working environment. The major regulations that were developed and updated (supported by a raft of other regulations that target specific areas of concern) are:

- Management of Health and Safety at Work Regulations 1999 (MHSWR).
- The Provision and Use of Work Equipment Regulations 1998 (PUWER).
- The Reporting of Injuries, Diseases and Dangerous Occurrences Regulations 1995 (RIDDOR).
- The Control of Substances Hazardous to Health (COSHH) Regulations 1994.

- Construction (Design and Management) (CDM) Regulations 1995.
- Control of Major Accident Hazards (COMAH) Regulations 1999.
- Pressure Systems Safety Regulations 2000 (PSSR).
- Dangerous Substances and Explosive Atmospheres Regulations 2002 (DSEAR).

Both the OSHA and the UK regulations place the onus on the employer to ensure that all work processes are safe and that the equipment used is safe. These requirements assume that designers are just working to supply plant and equipment to the employer's requirements with no duty to ensure safety. However, it became apparent that designs of plants and machinery needed to consider health and safety in their conception. The need to create a safe work place for construction workers also became important due to the large number of site accidents. The CDM Regulation was the first of the EU directives to ensure the design and construction of a safe work place with consideration for the safe construction, operation, and maintenance of the facility to be built.

1.3.1 Evolution of the law to include safety in design

Following the Piper Alpha accident it was established that *safety* must start at the conception and design phase of a project. The engineers that design the plant and machinery should also be held to account for their work. Accordingly, the EU in the development of their 'new-approach directives' has also embraced this concept. One of the first of the new directives to be issued based on this principle was the Machinery Directive. This placed a duty on the manufacturer to only design and manufacture safe machinery.

The arrival of the Machinery Directive was a great cause for concern. In the concept of designing a metal forming machine to be safe, it is possible to design all the controls and safety systems to be an integral part of the machine. In the case of process machinery, the working environment is part of the plant design and so it was even considered that the whole of a process plant could come under the Machinery Directive requirements. This of course was nonsense. However, with the advent of the Pressure Equipment Directive it became obvious that in many cases the machine is part of a pressure system and the safety of the machine is also dependent on the safety of the pressure system. The EU directives now ensure that the design engineers are responsible for ensuring safety in the design of plant and equipment. This completes the loop with the employer still responsible for ensuring a safe working environment. The law now covers all activities from design, manufacture, operational use, and maintenance of plant and equipment. The safety and welfare of the worker has to be taken into account at all stages of the process. The most important of these equipment design EU directives as enacted by UK regulations are as follows:

- Machinery Directive
  (Supply of Machinery Safety Regulations 1992, as amended 1994).
- Directive on Equipment and Protective Systems Intended for Use in Potentially Dangerous Atmospheres (known as the ATEX Directive)
  (Equipment and Protective Systems Intended for Use in Potentially Dangerous Atmospheres Regulations 1996 as amended 2001).
- Electromagnetic Compatibility Directive (known as the EMC Directive)
  (Electromagnetic Compatibility Regulations 1992).
- Low Voltage Directive
  [Electrical Equipment (Safety) Regulations 1994].
- Pressure Equipment Directive (PED)
  (Pressure Equipment Regulations, 1999/2001).

1.3.2   The duty of compliance
The person or entity that is held responsible is called the duty holder. Under the regulations some statutory duties are required. In some cases they are self-regulating or self-certifying with documentation to be made available on demand. In other cases there is a duty to inform the regulatory authority of actions taken. A list of the major regulations showing the duty holder and the statutory duty required is given in Table 1.1.

As given in the table, the law was intended to assign a clear demarcation of responsibilities to each of the entities, the user and the supplier. In the case of process plant it has not been possible. Designers, operators, and maintenance engineers each have a different prospective of the hazardous areas of their work. In the case of the COMAH Regulations for example, a safety case has to be presented by the owners but the application can only succeed if the process plant has been designed with adequate safety features. On the other hand, the owner has to ensure that he has the correct

Table 1.1 The regulations: duty holders and duties required

| Regulation | Duty holder | | Statutory duty |
	Owner	Maker/supplier	
MHSWR	X		Self-regulating
PUWER	X		Ditto
RIDDOR	X		Report accidents and near misses
COSHH	X		Self-regulating
CDM	X		Notify HSE and name planning supervisor as duty holder
COMAH	X		Submit a safety case to the HSE
PSSR	X		Appoint a competent person to produce a written scheme of examination
DSEAR subset of COMAH	X		Draw up an Explosion Protection Document
Machinery Directive		X	Self-certification
ATEX		X	Appoint a notified body
EMC		X	Self-certification to an EU standard, otherwise verification by a competent body or a type test by a notified body
Electrical Equipment (Safety)		X	Ditto
PED		X	Appointment of a notified body as required

organization and training provisions in place in accordance with the regulations. Safe and reliable operation depends on a joint input, as will be shown in the chapters that follow.

1.3.3 Enforcement of the law

In the UK this is in the hands of the Health and Safety Executive (HSE). They have powers of inspection and investigation. In the case of self-certification or self-regulation it is the responsibility of the duty holder to carry out all the actions as required by the applicable regulation. In the case of accidents that result in death, investigations by the HSE that establish non-compliance with regulations can result in a case for manslaughter with a jail sentence and/or fine if found guilty. Should a hazardous case of non-compliance be found this might also result in a closure order.

1.4   Codes and standards

The oil industry has its roots in Texas. The development of its processes and equipment has originated in the USA. Not surprisingly the codes and standards used in the design of plant and machinery have been from the American Petroleum Institute (API) and the American Society of Mechanical Engineers (ASME). These have been systematically developed by generations of engineers from the oil companies and equipment manufacturers with the objective of improving their reliability. Loss of containment can lead to fire and explosion with a large loss of capital. This loss can far outweigh the cost of any improvement in reliability and so critical equipment was not subject to any cost constraints. The standards are based on common agreement and to minimize the cost many requirements of the standards are optional at extra cost. Even so, many oil companies were prepared to pay for their own additional supplements at even greater expense.

API codes and standards are prescriptive, requiring features based on lessons learnt. They require many features that meet the required EU regulations. They do not ensure overall compliance as they are intended only for the design and manufacture of the equipment.

In the case of chemical plant and steam boilers, these have been developed in parallel to the USA and, in some ways overlap with the US ASME codes and standards. This has resulted in national codes and standards being established in the industrial nations of Europe. The German DIN standards, the UK British Standards and the Dutch standards have been in use for many decades. Then there are the associated materials standards. They all differ to some degree especially in the need for Charpy impact testing of materials. This has resulted in differences in the design and manufacture of equipment depending on the country of origin.

The Pressure Equipment Directive requires the use of harmonized standards and EU technical committees are now (2003) hard at work formulating the standards to ensure that the best features of the existing national standards are retained.

It is important to understand the relationship between the EU directives, the UK regulations, and the codes and standards. The directives and the regulations lay down the requirements of the law. Codes and standards on the other hand explain how the requirements should be carried out.

Chapter 2

The Law

2.1 Introduction

The objective of creating a single 'common' market in the European Economic Area (EEA) goes back to the EEC treaty (the Treaty of Rome) that originally established the Community.

Despite the elimination of tariff and quota restrictions between member states, the common market was not yet a reality in 1985. The free movement of goods was still partially impeded by technical barriers such as differing national product standards. A range of national restrictions blocked the growth of a free and competitive market for services.

Achieving the free movement of goods was the basis of creating a single European market. The ministers therefore agreed on a 'New Approach' to solve this problem for business. These 'new-approach directives' (Community laws) require member states to be responsible for ensuring the health and safety on their territory of their people and in particular, of workers; notably in relation to the risks arising out of the use of machinery and other equipment. They set out 'essential requirements' (e.g. for safety), written in general terms, that had to be met before products could be sold in the UK or anywhere else in the Community.

The Community also agreed a common commercial policy covering trade relations with non-EEA countries. In many areas (mainly trade in goods) the Community now has obligations under the General Agreement on Tariffs and Trade (GATT). For example, not to discriminate against the trade of other members of GATT and not to increase trade barriers without giving matching concessions in return. The purpose of GATT is to provide an accepted framework for the orderly conduct of international trade and to encourage the progressive opening of world markets.

The implications of the Community's approach are made clear by the following statement:

The single market will be of benefit to Community and non-Community countries alike by ensuring continuing economic growth. The internal market will not close in on itself – 1992 Europe will be a partner not 'Fortress Europe'. The internal market will be a decisive factor contributing to greater liberal-ization of trade on the basis of the GATT principles of reciprocal and mutually advantageous arrangements.

The new-approach directives are generally of two types, the first of which is aimed at a broad product range rather than specific products. In the second type, the detailed technical specifications are replaced by 'Essential Health and Safety Requirements (EHSRs)'. These describe the objectives to be achieved. In addition they refer to 'Transposed harmonized standards' as a means of compliance with the essential requirements.

This guide is concerned with those directives that are relevant to process plant, the most important being: the Machinery Directive, the Pressure Equipment Directive, the ATEX Directive, and the Construction (Design and Management) Regulations.

2.2   Construction (Design and Management) (CDM) Regulations

The CDM Regulations place duties on all those who can contribute to the health and safety of a construction project. Duties are placed upon clients, designers, and contractors, and the regulations create a new duty holder: the planning supervisor. They also introduce new documents: health and safety plans, and the health and safety file.

The regulations cover the installation, commissioning, mainte-nance, repair, or removal of mechanical, electrical, gas, compressed air, hydraulic, telecommunication, computer, or similar services that are normally part of a structure. A structure is defined to include fixed manufacturing plant that involves construction work over 2 metres in height (i.e. process plant).

The regulations also make a radical change in the relationship between the designer and construction.

In the past the designer was judged in common law to be only responsible for his design as a finished product. The safety of temporary structures and how the design was built was the responsibility of the building contractor. There was a clear-cut demarcation.

The CDM Regulations have demolished this demarcation and the designer is now held responsible for constructability.

These ideas are new to the building industry where the architect and builder held their separate positions as given above. In the case of the plant construction industry, however, there is a completely different position. In most cases the contract is for engineering, procurement, and construction (EPC). Engineering is usually a small fraction of the total cost and anything that will reduce construction cost has a big impact. Any worker injury, especially in a skilled trade, impacts productivity and many EPC contractors make strenuous efforts to achieve zero lost-time accidents. For the process industries the CDM Regulations only impose a more structured approach on existing practices. However, the CDM Regulations have specified roles for the different parties as defined below.

2.2.1   The client (on initiation of a project)
The client is required to:

- ensure that financial provision is made and time is allowed for safety requirements in the initial planning of a project;
- ensure that prior notice is given to the nominated authority (Health and Safety Executive in the UK) of the project if it is expected to last more than 30 days or involve more than 500 persons. It is the duty of the appointed planning supervisor to ensure that this notice is given;
- establish the site development requirements identifying any applicable hazards;
- produce the project conceptual design and issue this to the project team;
- appoint a planning supervisor;
- appoint a competent contractor as principal contractor.

2.2.2   The planning supervisor
The planning supervisor has to ensure that all safety regulations have been recognized and have been complied with at all stages of the project. The major outcome of the work is the production of the 'safety plan' and 'the safety file'.

Safety plan
The plan should show how it is proposed that all hazards are to be identified and the risk to health and safety lowered to an acceptable level throughout all stages of the project. It will include a list of

scheduled activities and procedures to be used from design through to construction and handover.

Safety file

The safety file is required to contain a record of all the as-built design features, including all the information on risks to health and safety that could arise from operations and maintenance, and the maintenance tasks needed for safe operation.

The file must be given to construction for updating and finally checked by the planning supervisor before being formally handed over to the client prior to operation.

2.2.3   The designer

The designer is required to identify any risks to health and safety in the design that could arise during construction, operation, or maintenance either from the materials used or the facilities provided. The design must include all reasonable and practical features to avoid these risks in accordance with the principle of safety integration. The designer must:

- make clients aware of their duties under the regulations;
- give due regard, in the design, to health and safety;
- provide adequate information, to those who need it, about the risks to health and safety of the design;
- co-operate with the planning supervisor and, where appropriate, other designers involved in the project.

Design is taken to mean all necessary drawings and documentation.

2.2.4   The client (as operator)

The client has to receive the safety file prior to handover of the completed project and has the duty of its safe keeping for future reference. The client has a duty to consult the file concerning any maintenance work or any subsequent alterations to the plant. This in effect overlaps with PUWER. Based on the safety file information the client will also need to develop work permits, and other procedures such as those required by the Control of Major Accident Hazards (COMAH) Regulations.

2.3 Dangerous Substances and Explosive Atmospheres Regulations (DSEAR) 2002

This is the implementation of Directive 98/24/EC on the protection of workers from chemical agents, CAD (Chemical Agents Directive) and Directive 99/92/EC concerning the ATEX (Explosive Atmospheres) 137 Directive. It overlaps with the CAD and COSHH regulations that are concerned with health, but DSEAR is concerned with safety.

DSEAR scope

DSEAR is concerned with safety and the reduction of risk of fires, explosions, and exothermic chemical reactions. Substances include: petrol, liquefied petroleum gas, paints, varnishes, and types of combustible and explosive dusts that may be produced by work processes. The regulations are applicable to all industrial and commercial premises ranging from petrochemical plant to school laboratories.

Main requirements

Employers and plant designers are required to:

- identify the location of any hazardous substance or the processing of any hazardous substances;
- carry out a risk assessment of the processing or handling of the substance;
- provide measures to eliminate or reduce as much as possible the risk;
- provide measures to deal with accidents and emergencies;
- provide information and training.

2.3.1 Hazardous area classification

There are many types of plant and equipment that process or use dangerous substances. To prevent fire and explosion, it is necessary to prevent ignition of the substance in the event of a release. At the design stage, it is usual to identify the areas where these releases could occur as 'hazardous areas'. Apart from ensuring that any naked flames are not in these areas, it will also be necessary to ensure that no electrical arcing can take place.

The two major internationally recognized codes of practice are API RP 500 issued by the American Petroleum Institute and IP code Part 15 issued by the Institute of Petroleum.

The definitions of IP code Part 15 would appear to have been adopted by the EEC ATEX 99/92 Directive and extended to include other industries that are subject to explosive dust clouds. See Table 2.1.

The area of a zone

There is no reason to believe that the rules that are so well established for determining the extent of hazardous zones for refineries are inappropriate. In some cases these rules could be considered to be over-cautious.

The DSEAR, however, are intended to be a catch-all to include many other situations in addition to refineries. Therefore the DSEAR require a risk assessment type of approach so that the extent of a hazardous zone must be based on the consideration of:

- release rate (the greater the rate, the larger the zone);
- lower explosion limit (LEL) (the higher the LEL, the less dilution is required);
- ventilation (both amount and availability, and predominant wind direction if relevant);
- relative density (is the zone predominantly above or below the release?);
- plant topography (e.g. are there any trenches or pits to trap gas).

Risk assessment

This is required for the design of a new plant or before the introduction of a new work process that involves the use of dangerous substances. The risk assessment must determine the probability of release of a dangerous substance, its ignition and the possible consequences (extent of damage to life and property). Based on this assessment the plant designer, or employer, must decide on the appropriate safety measures to be adopted.

Safety measures

Action is required in accordance with the safety hierarchy of:

Elimination

Avoid the use of the hazardous substance where possible.

Control measures

The hierarchy of control measures, consistent with the risk assessment and as appropriate, is to:

- reduce the quantity of the dangerous substance;

Table 2.1 API code and IP code classifications compared

API RP 500		IP code Part 15	
Class	Definition of location	Class	Definition of area
Class 1, Division 1	Ignitable concentrations of flammable gas are expected to exist or where faulty equipment might release gas and cause failure of electrical equipment	Zone 0	Where a flammable atmosphere is continuously present, or present for long periods
Class 1, Division 2	Ignitable concentrations of flammable gas are present, but are confined, or prevented from accumulation by adequate mechanical ventilation, or are adjacent to a Division 1 area from which gas could occasionally be communicated	Zone 1	Where a flammable atmosphere is likely to occur in normal operation
		Zone 2	Where a flammable atmosphere is not likely to occur in normal operation and, if it occurs, will only exist for a short period
	ATEX Directive extension		
		Zone 20	Where a flammable atmosphere in the form of a combustible dust cloud is continuously present, or present for long periods
		Zone 21	Where a flammable atmosphere in the form of a combustible dust cloud is likely to occur in normal operation
		Zone 22	Where a flammable atmosphere in the form of a combustible dust cloud is not likely to occur in normal operation and, if it occurs, will only exist for a short period

- avoid or minimize releases;
- control releases at source;
- prevent the formation of an explosive mixture;
- collect, contain, and remove releases to a safe place;
- avoid ignition sources;
- segregate incompatible substances.

Mitigation
The measures to be considered include:

- controlled access to reduce the numbers of people exposed;
- providing explosion-resisting features such as underground control rooms;
- providing explosion-suppression or explosion-relief equipment;
- providing means to control or minimize the spread of fires;
- providing suitable personnel protection equipment.

2.3.2 Risk management
Risks must be controlled by:

- design measures;
- maintenance of safety-critical items and the provision of adequate safety warning signs;
- development of work permits, operating procedures, and supervisory systems;
- instruction, training, and regular drills;
- emergency procedures and planning in accordance with the COMAH requirements as applicable to the situation.

Protective systems
The DSEAR require the designer or employer to be responsible for deciding on the type of protective system to be used. This must be based on the results of a risk assessment. Unfortunately the definitions of a protective system are given in the ATEX (Equipment Directive). This may lead to the erroneous impression that it is the responsibility of the equipment supplier. This is not so, the designer or employer must decide on the protective system to be used.

The process plant designer will need to be aware of the following definitions.

Equipment Group II, Category 1
Intended for use in a Zone 0 area classification (see below for zone classifications).

Equipment Group II, Category 2
Intended for use in a Zone 1 area classification.

Equipment Group II, Category 3
Intended for use in a Zone 2 area classification.

Equipment Group I, Categories M1 and M2
These follow the same definitions as given for Group II except that in the case of Category M2 equipment they are intended to be de-energized in the event of an explosive atmosphere.

The different characteristics of the above can be found below in Section 2.4.3.

From the above it would appear that the use of the equipment groups must always be used for the corresponding hazardous zones. This may well be so in the case of process plant. But it is not necessarily the intent. They must be selected on the basis of the risk assessment.

The continuous presence of a very small leak of a dangerous substance is classified as Zone 0. If it is in open air, easily dispersed and it is not easy to ignite, a lower category of protection could be justified. A hazardous area classified as Zone 1 in a building may warrant a higher level of protection.

A new departure is that area classification rules are to be extended so that mechanical machines will need to be certified in the same way as electrical machines. This also brings the potential need for retrospective certification for mechanical equipment used in flammable hazardous areas. This will also be needed where the electrical equipment has not been certified in accordance with ATEX Directive 94/9 as given below.

2.3.3 Required documentation
The directive requires the employer to draw up and keep up to date an 'explosion protection document'. Ideally this must be done during the design phase of a plant and certainly prior to operating the plant. The purpose of the document is to demonstrate in particular that:

- explosion risks have been determined and assessed;
- adequate measures will be taken to attain the aims of the directive, which is to ensure a safe and healthy working environment;
- work areas are classified into zones as applicable;

- all work places and work equipment, including warning devices, are designed, operated, and maintained with due regard for safety.

The document must be revised when the work place, work equipment, or organization of the work undergoes any significant changes, extensions, or modifications.

2.4 The Equipment and Protective Systems Intended for Use in Potentially Explosive Atmospheres Regulations 1996 (S. I. 1996/192), (ATEX Directive 94/9/EC, as amended 2001)

2.4.1 Introduction

This directive harmonizes the technical and legal requirements of such equipment and systems for use throughout the EU. Equipment includes electric motors, compressors, diesel engines, light fittings, control and communication devices, and monitoring systems. It also covers components that are essential for the safe function of equipment, protective systems, and detection equipment (including the parts which are located outside the hazardous area) that are intended to function as a whole.

In order to comply, equipment and systems are required to meet CEN, CENELEC, or BSI standards, or alternatively to meet the EHSRs of the directive.

2.4.2 The Essential Health and Safety Requirements (EHSRs)

Where approved standards are not available for mechanical equipment, then it must comply with the EHSRs summarized as follows:

Common requirements for equipment and protective systems
Section 1. General requirements
Dealing with general issues, material selection, design and construction, potential ignition sources, external effects, safety devices, and safety requirements.

Section 1.1 Selection of materials
The materials used must not of themselves trigger an explosion and they must be able to withstand all possible changes in the operating environment.

Section 1.2. Design and construction
Control of: potential ignition sources, surface temperatures, safe opening, dust ingress or egress.

Section 1.3. Potential ignition sources
Every possible source of ignition must be prevented; examples are: sparks, flames, electric arcs, hot surfaces, all forms of radiation energy, static electricity, stray currents, heat from rubbing surfaces, and energy from shock waves due to pressure control systems.

Section 1.4. Hazards arising from external effects
Operation of equipment must be reliable under all foreseeable changes in operating conditions and attack from aggressive substances.

Section 1.5. Requirements in respect of safety-related devices
Covering the need to ensure reliability, operability, and maintainability in preventing or minimizing the effect of any explosion under all conditions of operation. These requirements are to include software as well as hardware.

Section 1.6. Integration of safety requirements relating to the system
Concerning the need for emergency shut down and safeguards for power failure system interface errors, and to ensure the correct placement of detection devices.

In addition to the above, further EHSRs are applicable, depending on the equipment category: M1, M2, 1, 2, or 3.

2.4.3 Equipment groups and categories
The directive divides all equipment, including where necessary devices and components, into two groups.

- *Group I.* Comprises equipment intended for use in mines or the surface of mines where there is a possible risk of firedamp or combustible dust.
- *Group II.* Comprises equipment intended for use in other places likely to be at risk from explosive atmospheres.

The groups are in turn subdivided into categories. In the case of Group I they depend on the applicable factors, such as de-energizing in the event of an explosive atmosphere being detected.

In the case of Group II the applicable category depends on a risk assessment by the user of the likelihood and duration of an explosive atmosphere being present and the consequence of a fire or explosion.

The defining EHSR (there are many others) for each equipment category is as follows.

Category 1
Where an explosive atmosphere is present for long periods. The means of protection to be characterized by:

- either, in the event of failure of one means of protection at least an independent second means provides the same level of protection;
- or, the requisite level of protection is assured in the event of two faults occurring independently of each other.

Could be described as 'Safe even with rare malfunctions'.

Category 2
Where an explosive atmosphere is likely to occur during normal operation:

- the level of protection to be ensured, even in the event of frequently occurring disturbances or faults that normally have to be taken into account.

Could be described as 'Safe with normally expected malfunctions'.

Category 3
Where an explosive atmosphere is only likely under abnormal circumstances:

- the level of protection to be ensured during normal operation.

Could be described as 'Safe in normal operation'.

Category M1
Mining equipment that can remain energized in the presence of an explosive atmosphere. The same characteristics as Category 1.

Category M2
Mining equipment that must be de-energized when an atmosphere exceeds the lower explosion limit. Otherwise, the same characteristics as Category 2 apply.

2.4.4 *Conformity assessment requirements*
For:

- *Categories 2 and M2*. Non-electrical equipment.
- *Category 3 equipment*.

A manufacturer's internal assessment is required. The technical file, except for Category 3 items, must be deposited with a notified body.

For:

- *Categories 1 and M1.* Protective systems.
- *Categories 2 and M2.* Electrical equipment and internal combustion engines.

An EC-type examination by a notified body is required.

It is suggested that in all cases a notified body should be engaged to verify compliance and issue a certificate of conformity.

Definition of a notified body

The notified body is responsible for carrying out testing and conformity assessment of the design and product verification (routine auditing) of subsequent manufacture or alternatively production quality assurance (QA) (auditing of the manufacturer's ISO 9002 quality control system) as applicable.

The notified body means a national certification authority appointed by the national government as notified to the EC. The notified body for the UK is the Electrical Equipment Certification Service (EECS) otherwise known as BASEEFA.

It should be noted that there is a reciprocal certification agreement between the EECS and Factory Mutual of the USA.

Documentation and marking

A technical file is required in all cases as a record of the measures taken in compliance. The list of contents is the same for all directives and is described in Section 2.6.5.

Marking is required as follows:

- the well-established EX sign in a hexagon is required together with the symbol of the equipment group and category;
- for Group II equipment, the letter 'G' (for gas) and/or the letter 'D' (for dust);
- the name and address of the manufacturer;
- series or type identity, serial number, etc.

2.4.5 Mechanical equipment

The EHSRs that affect mechanical equipment will be those with regard to hot surfaces and potential ignition sources. Generally speaking most process machinery will have limited discharge temperatures for mechanical reasons, so maximum casing temperatures are likely to be less than 200 °C. In the case of a centrifugal compressor, for example, the possible ignition sources are:

(a) Sparks due to the coupling guard touching the coupling. Two methods of control are used. First to ensure the rigidity of the guard to prevent contact, and second to make it of non-sparking material in case it should come into contact. This is a rare happening and not likely to occur. In the past this was usually done for Divison 1 areas, see Table 2.1.

(b) A hot surface due to a bearing overheating. This can be considered to be an expected failure that is usually safeguarded by monitoring its temperature. The use of two thermocouples will provide redundancy. This is normal to avoid bearing failure and machine damage.

In order to comply with the ATEX regulations, safeguards (a) and (b) would be needed for Category 1 equipment and only safeguard (b) for Category 2 equipment.

2.5 Provision and Use of Work Equipment Regulations (PUWER) 1992

These regulations lay down important health and safety laws for the provision and use of work equipment. Information on these regulations is covered in HSE Document L22 – Guidance on Regulations.

The term 'work equipment' means any machinery, appliance, apparatus, or tool and any assembly of components which, in order to achieve a common end, are arranged and controlled to function as a whole. The requirements imposed on an employer by this legislation apply to work equipment provided for use, or used by, any of the employees.

In summary the regulations require that equipment provided for use in the work place needs to be:

- selected to be both safe and suitable for the task;
- maintained in a safe condition;
- inspected to ensure safety, with quality assurance records;
- only used by, and accessible to, qualified persons who have received adequate information, instruction, and training;
- equipped with suitable safety measures such as controls, protective devices, markings, and warnings signs, etc.;

- generally in conformity with any other related health and safety regulations that are applicable to the place of work.

There are also specific requirements that concern mobile work equipment, power presses, and miscellaneous other equipment. These and all other machinery must be supplied in accordance with the requirements of the Machinery Directive.

2.6 Machinery Directive {[Supply of Machinery (Safety) Regulations] 1992, as amended 1994}

2.6.1 Introduction

The Machinery Directive is one of the new-approach directives and is implemented in the UK by the Supply of Machinery (Safety) Regulations 1992, as amended 1994.

All 'new machinery' supplied in the EEA must comply with the Machinery Directive (89/392/EEC). The regulations make it an offence for a responsible person to supply machinery unless it is safe. To be safe it must comply with the requirements of the directive. The responsible person is defined as the manufacturer or his appointed agent. Compliance is to be indicated by applying the CE mark to each machine. Some machinery will require type examination before it can receive the CE mark. Any products legitimately bearing the CE mark can be supplied without hindrance within the EEA.

Scope of the regulations

The regulations are intended to cover the supply of all new machinery and safety components for machines.

Any new model, either a one-off or for series production, must comply with the regulations. In addition to this, three other groups of machines are considered to be new, viz.:

(a) Second-hand machines imported from outside (and never previously used in) the EU.
(b) Newly supplied examples of machines, which have been supplied previously prior to the regulations.
(c) Some refurbished machinery where the performance is improved from its original level.

The last item requires some interpretation:

- refurbished machinery to be used for a different purpose is considered new;

- refurbished machinery to be used for the same purpose, and to the same specification, is not considered new, as this work is routine maintenance;
- refurbished machinery with any sort of modification has to be reviewed, see Section 2.6.9.

In order to comply with the regulations, all machines must comply with the applicable Essential Health and Safety Requirements (EHSRs).

The underlying principle of the EHSRs is the requirement for safety integration. This means identifying and assessing the risks posed by the machine and eliminating them by good design rather than tacking on a proliferation of guards and safety devices. This may not always be possible but the designer will have to demonstrate that all reasonable and practical measures were taken. The directive classifies machines into different groups as follows:

Group 1. Machines in general.
Group 2. Machines for making food, wood-working and hand-held machinery.
Group 3. Machines designed to have mobility.
Group 4. Machinery involved in lifting operations.
Group 5. Machinery intended for underground operations.
Group 6. Machines designed to move or lift people.

An important feature of the regulations is that some machinery, which is considered to pose 'special hazards', has to follow a different route to conformity assessment. The classes of machinery involved are listed in Schedule 4 of the regulations. They are specific types that generally fall within Groups 2 to 6 and depend on intimate human operation.

All machines need to comply with Group 1 EHSRs and there are additional EHSRs required for each of the other groups.

2.6.2 Individual machines and safety components

The basic definition of machinery is given as:

(a) An assembly of linked parts or components, at least one of which moves under power.
(b) An assembly of machines linked together to achieve the same end.
(c) Interchangeable equipment modifying the function of a machine.

Safety components for machinery are described as:

(d) Components which are supplied separately to fulfil a safety
    function when in use and the failure or malfunctioning of which
    will endanger the safety or health of exposed persons.

The intention of the directive is to have a definition that agrees with
the commonly understood concept of a machine, as a product ready
for use, unless Articles 1.3, 1.4, or 1.5 of the directive specifically
exclude it.

2.6.3 *Assemblies of machines*

The second paragraph of Article 1.2 of the directive deals with
assemblies of relevant machines which in order to achieve the same
end, i.e. a specific application of the kind envisaged in the first
paragraph, are arranged and controlled so that they function as an
integrated whole.

Examples of what is intended would be vehicle assembly lines or
assemblies in bottling plants and paper mills.

Individual machines within the assembly, which meet the definition
of machinery in the first paragraph of Article 1.2 of the directive, and
which are able to function independently, should be CE marked by
their supplier, who is responsible for fulfilling all the requirements of
the directive in relation to that individual product. The assembler of
the machines is responsible for:

- selecting suitable products to make up the assembly;
- putting together the assembly in such a way that it complies with
  the provisions of the directive;
- fulfilling all the requirements of the directive in relation to the
  assembly;
- CE marking the assembled whole.

There are, however, two exclusions in Article 1.3 of the directive that
should be noted:

(a) steam boilers, tanks, and pressure vessels;
(b) storage tanks and pipelines for petrol, diesel fuel, inflammable
    liquids, and dangerous substances.

Steam boilers, piping, and pressure vessels all come within the scope
of the Pressure Equipment Directive. While these items can be
considered separate to and not for incorporation with machinery, in
many cases machines are supplied for incorporating with a pressure
system.

The examples that follow illustrate how these requirements can be interpreted.

(a) *This means that a pump can be incorporated as part of another machine, such as a gas turbine. They are both machines and the documentation can be combined, i.e. incorporated.*

(b) *A pump cannot be incorporated as part of a boiler. A feed-water pump supplied as part of a boiler package must have its own documentation and be CE marked in accordance with the Machinery Directive.*

(c) *A boiler is a boiler subject to its own regulations and cannot be incorporated as part of a feed-water pump as defined by the Machinery Directive.*

(d) *A compressor package can consist of a compressor, various pressure vessels, pumps, and electric motors. While the pumps and the electric motors can be incorporated as part of the compressor, the pressure vessels cannot. The pressure vessels and piping systems will need to be in compliance with the Pressure Equipment Regulations (PER). The packager, however, under the Construction (Design and Management) Regulations, will be held responsible for the safety of the whole system.*

(e) *A process compressor package cannot function and be safe unless there are anti-surge controls, emergency shutdown facilities such as isolation and blow-down systems, etc. Gas detectors may also need to be installed, especially where the process gas contains H_2S, which is a dangerous substance that is toxic, flammable, and explosive. In these cases it may be more appropriate to issue a certificate of incorporation to the process plant contractor. The plant contractor is then responsible for the design of the whole installation. Compliance with the Pressure Equipment Directive, the ATEX Directive, and the Machinery Directive would be required.*

2.6.4 Machinery for incorporation

It is considered that 'machinery for incorporation' should satisfy the following:

(a) It meets the definition of machinery in Article 1.2 including being joined together for a specific application.

(b) While it can function independently, it is not envisaged by the supplier that it is ready for end use, because it is designed solely to be incorporated into, or assembled with, other products which, taken as a whole, will also meet the definition of machinery in the first or second paragraphs of Article 1.2 of the directive. Or it may be that it does not meet the EHSRs of the directive until assembled with other products such as a pressure system.

(c) The responsible person may, at his option, make a *declaration of incorporation*. The machinery is not then CE marked until it has been incorporated or assembled with other machinery or products. The declaration of incorporation is required to state specifically that the equipment to which it refers may not be put into service until the assembly has been declared to be in conformity with the Machinery Directive and its EHSRs.

2.6.5 Technical file

Objective of the technical file

(a) In some directives the technical file is the principal means of assessment of the conformity of a product within the framework of the market surveillance by the member states. For the Machinery Directive, the assessment of conformity is based almost exclusively on the manufacturer's declaration of conformity, without the intervention of a third party or a notified body.

(b) The file compiled by the manufacturer is intended essentially for the national inspection authorities who have the right to ask the manufacturers or the importer to communicate the data relating to the tests carried out concerning safety etc., when they have good grounds for believing that a product does not offer the degree of safety required in all respects. Refusal on the part of the manufacturer or the importer to communicate these data constitutes sufficient reason to doubt the presumption of conformity.

(c) It must, therefore, be possible to place this technical file at the disposal of the competent national authorities, should they so request, when the product is placed on the Community market.

(d) In other directives, the documentation or technical file is just one of the means of completing a specific conformity assessment procedure with the intervention of a third party (notified body). This is the case with the directives that provide only for the EC-type examination. This certificate is in turn included in the technical file.

In the case of machinery, the purpose of the technical file is to provide evidence to the authorities, when necessary, that the requirements of the directive have been met.

Format and contents of the technical file

The details included in the technical file always depend on the nature of the product and on what is necessary, from a technical point of view, to demonstrate the conformity of the product, either to the harmonized standards, if the manufacturer has followed them, or to the essential requirements of the relevant directive if the manufacturer has followed none, or only some, of the harmonized standards. This must, therefore, be determined case-by-case depending on the product.

The file itself need not permanently exist, but if required, it must be possible to assemble it and make it available in a reasonably short time. The responsible person has to ensure that the following information is available, as the content of the technical file, when required.

- An overall drawing of the machine or safety component, and drawings of control circuits.
- Full detail drawings, calculations, and test results, etc. that will enable the conformity with the EHSRs to be checked.
- A list of:
  - the EHSRs;
  - transposed harmonized standards (such as BSI);
  - standards; and other technical specifications that were used when the machinery or safety component was designed.
- A description of the methods adopted to eliminate hazards.
- As applicable, any technical report certificate obtained from a competent body or laboratory per EN 45000 or BS 75000.
- A copy of the user instructions.
- In the case of series manufacture, the quality control (QC) measures to ensure that the machinery remains in compliance.
- The results of tests by the manufacturer to prove that the machinery or safety component is capable of being erected and put into service safely.

Availability of the technical file

The technical file must be available to the national authorities for inspection and control purposes. With certain exceptions, this obligation to make available a technical file inside the territory of the Community starts at the time of the placing of the product on the Community market, whatever the geographical origin of the product. This obligation is incumbent upon the manufacturer or their representative established in the Community. If the manufacturer is not established in the Community and has no representative in the Community, the person who places the product on the Community market must take on this obligation.

Any person responsible for placing a product on the Community market, but not in possession of the technical file, must be capable of:

● stating where the technical file is situated inside the Community;
● presenting the technical file as soon as possible on request from the national authorities.

However, the name and address of the person in possession of the file need not be expressly mentioned on the product or on its packaging, unless otherwise specified.

The file can be requested only during checks made for market surveillance purposes by the member states. The full technical file can only reasonably be requested where necessary, and certainly not when only an individual point is to be checked, in which case only the relevant part of the file should be required.

Community-wide organization of the market surveillance procedures and co-ordination of the inspections should avoid repeated submission of the same technical file by the same manufacturer to different inspection authorities.

The data for the technical file must be kept for at least 10 years from the last date of manufacture of the product, unless the directive expressly provides for any other duration.

Language of the technical file

The manufacturer or their authorized agent in the EU must draw up the file in one of the EU languages. A member state may request presentation of the technical file in its official language, but should not do so if the national authorities can understand the file or its contents in the other language. Where a translation is required, the person in possession of the file will be allowed extra time to submit the file to the inspection authorities. Moreover, no further conditions may be imposed concerning this translation, such as a requirement of a translator accredited or recognized by the public authorities, or of official translators or other similar requirements.

It should be noted that UK regulations require that if the file is drawn up in the UK then English is to be used.

2.6.6 Confidentiality

Directive 90/683/EEC (Annex 1.1) stresses the need to ensure the legal protection of confidential information. No exceptions can be made to this very important principle, which the member states must observe strictly. To this end, member states must ensure that everyone involved in the assessments, inspections, and surveillance, and who has knowledge of the contents of the technical file is bound to professional secrecy. Precise rules will, where necessary, have to be laid down by the member states to guarantee this confidentiality. This applies in particular to the bodies notified by the member states, who must ensure that these bodies maintain this confidentiality.

Confidentiality is also mentioned in the EN 45000 series of standards that serve as the reference standards for the notification of bodies by the member states.

2.6.7 Machinery modifications

The Machinery Directive requires that any modification must undergo the same procedure as for a new machine. In addition to design alterations, some changes in manufacturing arrangements and documentation should also be dealt with as modifications. Examples of this type include:

- significant changes to the quality system;
- revision of the hazard/risk assessment or other major sections of the technical file;
- for 'authorized representatives', changes in the relationship with the manufacturer.

Modifications within the requirements of the Machinery Directive are those that have an effect on safety. There is no simple definition of those modifications that should be regarded as safety related, but in many instances this will be self-evident. The Machinery Directive implicitly requires a hazard/risk assessment of changes to a machine; carrying out this assessment should determine whether or not the modification has any potential safety significance.

It follows that an assessment must be carried out for any modification however small. See Section 6.6.

If the assessment shows no impact on safety then a record of the assessment and technical details of the modification should be filed.

For example, a pump with material changes, such as a shaft in some new material, could very well have no impact on safety.

If there is an impact on safety, then:

- an old machine with no CE mark will require a technical file in accordance with the Machinery Directive and a CE mark must be applied;
- any existing CE mark on a machine will no longer be valid and a new one must be applied. The technical file must be updated in accordance with the Machinery Directive.

Like for like changes will usually have no impact on safety. Where a change of technology is involved there usually is an impact. Changing a pneumatic control system to an electronic one could be said to fulfil the same function. The failure modes will be different and safety could be affected.

2.6.8 Documentation

A 'declaration of conformity' has to be issued by the responsible person for each machine. The declaration should include the following information:

- the name and address of the responsible person;
- the full address of the manufacturer;
- description of the machine, type, and serial number;
- all directives with which the machine complies and, as appropriate, a statement of conformity with the EHSRs;
- a list of the harmonized standards and national standards used;
- identification of the signatory.

If the machine is supplied for incorporation, then a 'declaration of incorporation' has to be issued instead of a declaration of conformity. The content required is the same but with the addition of a statement that the machine must not be put into service until the final installation has been declared to be in conformity with the provisions of the directive and all applicable EHSRs.

2.6.9 Summary of actions required by the regulations and amendments

The supplier of 'relevant machinery' must ensure that:

(a) The Essential Health and Safety Requirements as listed in Schedule 3 to the regulations and as amended are satisfied.

(b) A conformity assessment has been carried out by the responsible person.
(c) The responsible person has issued a declaration of conformity, or of incorporation as appropriate.
(d) A CE mark has been affixed unless a declaration of incorporation has been issued.
(e) The relevant machinery is in fact safe.
(f) The data required for a technical file have been identified and are made available for 10 years.

Any modification requires a review of all the work previously carried out. Any impact on safety requires a new or revised technical file to be prepared as appropriate. A new CE mark must be applied.

2.6.10 Machinery for process plant

The Machinery Directive has been drafted to cover a whole range of machinery. It is intended to safeguard people from fatal accidents and injury. It must be understood that wood-working machinery and machines for sheet metal work, etc. require constant human attention. There is a high risk of injury and quite complex safety systems are needed.

Process machinery such as pumps, compressors, and their drivers are run with little human intervention. The main hazard is the loss of containment of a dangerous substance. The only exposed running part is the coupling, which of course has a guard. A health hazard is noise, and injury could be caused by a hot surface.

The highest risk to people is in maintenance operations, lifting of heavy items, the need to lock out drivers to prevent start-up, venting, purging, and isolation from the process. Reciprocating compressors are especially dangerous, as any inadvertent crankshaft movement during maintenance will result in fatalities if people are inside.

Process machinery such as pumps, compressors, and turbines will normally only need to comply with the EHSRs listed under Group 1. A summary of these is given in Appendix C in the form of a checklist.

Most of these issues are covered in API and major oil company specifications and the EHSRs of the Machinery Directive should not result in any additional work except for the issue of the relevant declarations. The technical file is not required but the data for its assembly have to be available. In the event of an accident involving a machine, this documentation can be proof that due care for safety had been exercised.

In practice, the decision as to whether the machine should be supplied CE marked or with a certificate of conformity will depend on the circumstances. Standard off-the-shelf machines will normally be CE marked. Machines that are made to a customer specification will more likely be supplied with a certificate of conformity. In any event this will need to be a matter for agreement between the purchaser and the vendor. In these cases the technical documentation required is usually specified in the purchase order as will the degree of third-party inspection and verification of the EHSRs.

Interpretation of the legal requirements for a CE mark

(a) Purchase orders for machinery or complete assemblies of machinery should require the supplier to issue a declaration of conformity and to CE-mark the complete assembly.

(b) Where the user or contractor is the assembler of the machinery: they should ensure compliance with the essential requirements, and compile and retain a technical file; they must issue a declaration of conformity and affix a CE mark; exceptionally, they should ensure that they are meeting the special requirements for Schedule 4 hazardous machinery.

(c) Where the user or contractor is purchasing machinery for incorporation into an assembly for which they will be taking overall responsibility, they should ensure that the suppliers of the components issue a declaration of incorporation and acknowledge that they are retaining the technical file required by the regulations. This to be made available on request by the user or contractor or by the HSE (or equivalent enforcement authority in other member states of the EU).

(d) Purchase orders should explicitly state the user or contractor's requirements regarding recommendations (a) to (c) above.

It should be noted that machinery could be subjected to a number of different directives. The one CE mark is intended to signify compliance with all other specific directives. These must also be stated on the declaration of conformity or the declaration of incorporation as applicable.

2.7   Electromagnetic Compatibility (EMC) Regulations 1992

The EMC Regulations require that apparatus shall be so constructed that:

- the electromagnetic disturbance it generates does not exceed a level allowing radio and telecommunications equipment and other relevant apparatus to operate as intended;
- it has a level of intrinsic immunity that is adequate to enable it to operate as intended when it is properly installed and maintained, and used for the purpose intended.

As an example, a software process control system must not be capable of being affected to the extent of preventing it from operating as intended, for example, as a result of electromagnetic interference from say a fluorescent light. Neither must the system itself cause any other equipment to be affected by the emission of electromagnetic radiation.

For the EMC Regulations there are three routes to compliance.

(a) The *standards route*. This is available for apparatus other than radio communication transmitting apparatus. It involves self-certification by the manufacturer against the appropriate harmonized European standards.

(b) The *technical construction file route*. This is available for apparatus other than radio communication transmitting apparatus, where there are no European harmonized standards or where the manufacturer chooses not to apply such standards. The technical construction file is required to include a report or certificate issued by a 'competent body' appointed by the Department of Trade and Industry (DTI).

(c) The *EC type-examination route*. This is applicable for radio communication transmitting apparatus including mobile telephones. An EC type-examination certificate is required to be issued by a 'notified body' appointed by the DTI.

2.8 The Electrical Equipment (Safety) Regulations 1994

This directive, implemented by the Electrical Equipment (Safety) Regulations 1994, requires that 'electrical equipment must comply with the requirements of Regulation 5(i)'. That is to say it must be:

- safe;
- constructed in accordance with principles generally accepted within member states as constituting good engineering practice in relation to safety matters;

- designed and constructed with an adequate level of protection from electric shock in accordance with the relevant harmonized standards such as issued by the BSI;
- in conformity with the principle elements of the safety objectives of Schedule 3 of the regulations. These are:
  (i) Generally to ensure that instructions are provided to ensure that the equipment is used safely. The equipment must be made to ensure that it can be safely and properly assembled and connected. The manufacturer's name or trademark must be clearly displayed. Equipment must be designed and made to ensure protection from the risks given in (ii) and (iii).
  (ii) To provide protection against hazards arising from the electrical equipment.
  (iii) To provide protection from hazards which may be caused by external influences on the electrical equipment.

Products affected

All electrical equipment intended for use between 50 and 1000 V AC or 75–1500 V DC. The regulations are not applicable to specialized products that are covered by other regulations such as ATEX, medical equipment, etc.

Compliance

In order to comply, equipment and systems are required to meet CEN, CENELEC, and BSI standards or to meet the Safety Requirements of the directive. If the product is not in conformity with a recognized standard but has been designed and built to the Safety Requirements of the regulations then verification by a notified body is desirable.

Documentation requirements

The manufacturer or his appointed agent must ensure:

- the affixing of a CE mark;
- the drawing up of a declaration of conformity;
- the compiling and holding of the technical documentation.

In some cases of equipment built to customer order it could be appropriate to issue a declaration of incorporation.

2.9   Pressure Equipment Directive (PED)

2.9.1   Introduction

This directive is implemented by the Pressure Equipment Regulations 1999 (SI 1999/2001).

The directive is intended to regulate the design, manufacture, and QC of pressure equipment to ensure that they are safe. The directive applies to all new pressure equipment sold in the EU. The directive also applies to second-hand equipment imported from outside the EU.

Scope

Pressure equipment includes boilers, vessels, piping, safety accessories, pressure accessories, and their assemblies. All components that go towards the make up of a pressure system are included. Any subsequent modifications to pressure equipment are also included. A responsible person has to ensure that pressure equipment and assemblies above specified pressure volume thresholds comply and, that:

- they are safe;
- they meet EHSRs covering design, manufacture and testing;
- they comply with applicable conformity assessment procedures;
- they carry the CE marking and other required information.

Pressure equipment and assemblies that fall below the pressure of 0.5 bar and the specified volume or pipe size are only required to be safe, designed and manufactured to sound engineering practice, and to carry the normally specified markings. Sound engineering practice means whatever standards and materials that are established and normally used.

Exclusions

Nuclear plant and transmission pipelines are excluded, as are items that are regulated by other directives such as the Machinery Directive. Any equipment bought by a user is included, but any fabrication on site by the user, such as pipework, is excluded. Work carried out by the user on site for use by the user will be covered by other regulations such as PUWER and the PSSR.

The responsible person

The manufacturer, if located in the EU, or their nominated agent if the manufacturer is outside the EU, must ensure that any pressure equipment sold complies with the requirements of the directive.

The PED covers any pressure equipment or system that is intended to function as a whole. This means the steam system with the connected items of a boiler, or the whole of a refinery or process plant. In this situation the responsible person is defined as the person with overall responsibility for the design, manufacture, and QC of the plant process systems. This is usually an engineering, procurement, and construction (EPC) contractor. The EPC contractor will subcontract pipework fabrication and purchase other pressure equipment to their design and specification. In this case the EPC contractor will be responsible for a 'global conformity assessment' of the whole plant to ensure compliance with the regulations. A notified body (third-party quality assurance, QA) will be needed for design checking and surveillance of the subcontracted fabrication activity to enable CE marking of the equipment. In other cases the user may choose to be the responsible person, in which case the user has to employ the notified body and they together have to ensure compliance with the regulations.

Notified bodies

The use of notified bodies is in most cases part of the compliance assessment process. It is the duty of the manufacturer (the responsible person) to select and appoint one. Notified bodies are accredited by the United Kingdom Accreditation Service (UKAS), a department of the DTI. There are national bodies that fulfil the same functions for each member state of the EU, such as SOCFRAC, AENOR, etc. Notified bodies will be the same third-party inspection insurance organizations that were used prior to the PED in a new guise. They are responsible for the checking of designs, approval of welding procedures, welders, and non-destructive testing (NDT) and personnel. They are also responsible for the accredited certification of manufacturers' own quality systems to the ISO 9000 series.

2.9.2 Harmonized standards

Harmonized standards have been developed to be used throughout the EU. They are intended to replace the many differing national standards that have been in common use in the past. The major product standards are:

- BS EN 13445 series for unfired pressure vessels.
- BS EN 12952 series for water tube boilers.
- BS EN 13480 for industrial piping.

There are many more standards covering all items of pressure equipment, components, and materials. All aspects of design, manufacture, and inspection are also covered. The advantage of using the harmonized standards is that compliance with all or some of the EHSRs can be presumed.

2.9.3   The EHSRs

The directive places full responsibility on the manufacturer to comply with the Essential Health and Safety Requirements. These are summarized as follows.

(1) General.
   Must be designed to be safe for all foreseeable conditions of operation and designed to prevent misuse or give warning to prevent misuse.
(2) Design.
   - *Adequate strength.* Adequate for internal–external pressure, ambient and operating temperatures, static pressure and mass of contents in operating and test conditions, traffic, wind and earthquake conditions, reaction forces and moments from pipes, etc. Design life relative to creep, erosion and corrosion, fatigue, and the effect of the decomposition of unstable fluids. Measures taken to ensure structural stability, especially during transportation and handling.
   - *Provisions to ensure safe handling and operation.* Openings and closures, dangerous pressure blow-off from vents and safety valves, devices to prevent entry when under pressure/vacuum, protection from harmful surface temperatures and decomposition of unstable fluids.
   - *Means of examination.* Provision of manholes and inspection holes as appropriate.
   - *Means of draining and venting.* To avoid harmful effects, and for all stages of operation such as maintenance, pressure testing, etc.
   - *Corrosion and chemical attack.* Adequate allowance or protection for intended and foreseeable use.

- *Wear*. Design for erosion/abrasion, material allowance and/or use of cladding materials. Operator instructions for measures needed to ensure safe use.
- *Assemblies*. Must be suitable and reliable for the duty.
- *Provisions for filling and discharge*. To ensure safe operation, to ensure safe connection/disconnection.
- *Protection against over-pressure*. Provision of safety devices, indicators, and alarms.
- *Safety accessories*. Provide dedicated and reliable pressure-limiting devices, and temperature monitoring devices. The designs must adopt the principles of fail-safe modes, redundancy, diversity, and self-diagnosis.
- *External fire*. Facilities for damage limitation in the event of fire.

(3) Manufacturing.

- *Manufacturing procedures*. Details the need to control quality by the use of NDT, heat treatment, welding procedures, welder qualifications, QC/QA, and use of notified bodies, etc.
- *Final assessment*. The verification, by a notified body, of QC/QA documents for compliance with the PED, final inspection, and hydrostatic proof test. In the case of assemblies, safety devices must also be included.
- *Marking and labelling*. The need for a data plate giving all technical details of the design parameters so as to prevent misuse. Warning labels against misuse where applicable.
- *Operating instructions*. To include all relevant safety information with regard to mounting, putting into service, use, maintenance, maintenance checks and inspection, and any warnings with regard to misuse.
- *Materials*. To be in accordance with the design code and the material specification, suitable for the duty, including all welding materials.

2.9.4 Conformity assessment procedure

The regulations require all pressure equipment to be classified in accordance with the potential risk. This depends on the type of fluid and the pressure and volume of stored energy. The conformity assessment is the verification that the pressure equipment has been designed and manufactured in accordance with the regulations. The degree of verification depends on the potential risk. On completion of the conformity assessment procedure the manufacturer is required to issue a certificate with a declaration of compliance and apply a CE mark. In the case of equipment for incorporation, the manufacturer must issue a declaration of incorporation. The responsible person for the global compliance assessment will be required to CE mark the final assembly. See Section 2.6.8

Classification steps

(1) The contained fluid is classified as 'gas' or 'liquid'. A fluid that has a vapour pressure of greater than 0.5 bar above atmosphere when at the maximum allowable temperature of the equipment is to be classified as a gas. Any release will then become gas. Conversely a fluid with a vapour of 0.5 bar or less under the same conditions as above can be classified as a liquid as it will not vaporize when released.

(2) The fluid is classified further as being:
   • Group I, a dangerous substance.
   • Group II, all other fluids including steam.
     This then results in four categories of fluids: gas Groups I and II, etc.

(3) The next step is to classify the type of pressure equipment in which the fluid is contained: vessel, steam generator, piping (pressure accessories in effect can be classified as piping).

(4) The equipment is then classified according to their pressure and volume or pipe size in accordance with the nine classification charts: four each for piping and vessels (one for each fluid classification) and one for steam generators.

(5) From the applicable classification chart the equipment is then classified as being Category I, II, III, or IV. When it falls outside these categories, the equipment should be of sound engineering design. Sound engineering design means conforming to any recognized code or standard normally in use prior to the PED. It should be noted that safety accessories are generally considered to be Category IV equipment unless they are made to protect specific pressure equipment of which they will then form a part.

Study of the charts will show that the pressure volume or size boundaries differ depending on the fluid. A much larger vessel can be designed and fabricated without the need for a notified body (third party) intervention for a Group II 'safe' liquid compared with that for a Group I 'dangerous' gas. The vessel categories are therefore determined by the $PS \cdot V$ (pressure in bars $\times$ volume in litres) boundaries, which change depending on the fluid contained. Those for piping are similar except they are determined by PS·DN (pressure in bars × DN pipe size in millimetres) boundaries. Category IV equipment is considered to be that with the highest risk especially when in service with Group I gas.

Conformity assessment modules

There are 13 different conformity assessment modules, the selection of which will depend on the equipment classification and, with certain modules, the certification held by the manufacturer's QC organization. This is reflected in the choice of modules available for selection under Category II, III, and IV equipment. Note also that for Category III and IV equipment double modules are required. See Table 2.2.

The modules are graded from A with no involvement of a notified body, to H the highest grade with the maximum involvement. Manufacturers are free to choose a higher grade should they find it more convenient.

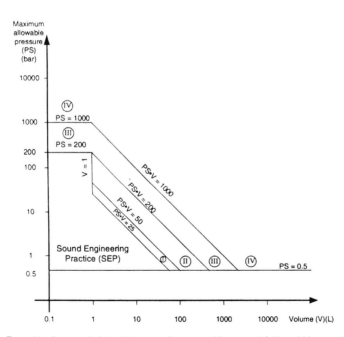

Exceptionally, vessels intended to contain an unstable gas and falling within categories I or II on the basis of Chart 1 must be classified in category III

Chart 1

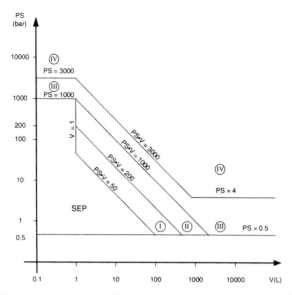

Exceptionally, portable extinguishers and bottles for breathing equipment must be classified at least in category III

Chart 2   Vessels for Group 2 gases

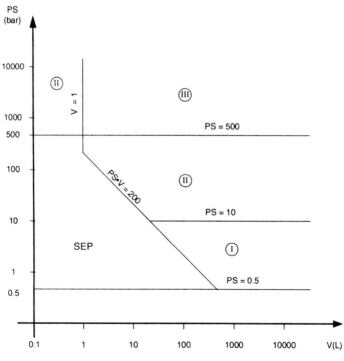

Chart 3   Vessels for Group 1 liquids

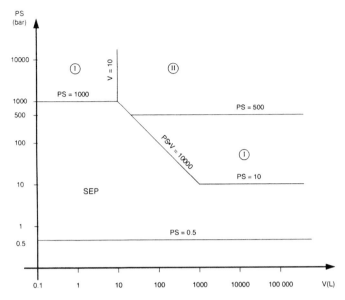

Exceptionally, assemblies intended for generating warm water at temperatures not greater than 110 °C which are manually fed with solid fuels and have a product of pressure and volume greater than 50 bar litres, must be subject either to an EC design examination (Module B1) with respect to their conformity with Sections 2.10, 2.11, 3.4, 5(a) and 5(d) of the essential safety requirements, or to full quality assurance (Module H).

Chart 4   Vessels for Group 2 liquids

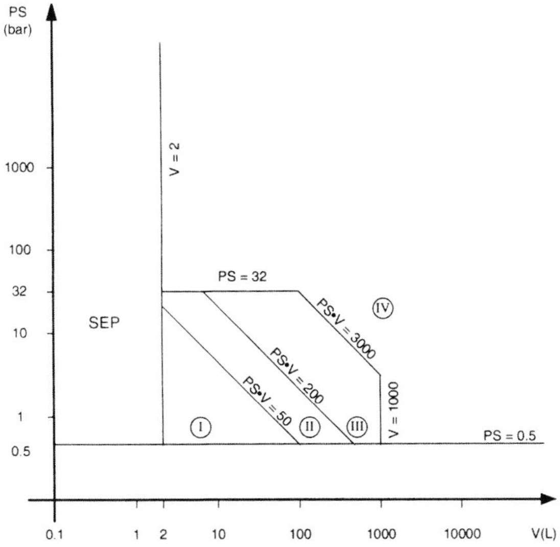

Exceptionally, the design of pressure cookers must be subject to a conformity assessment procedure equivalent to at least one of the category III modules.

Chart 5   Steam generators

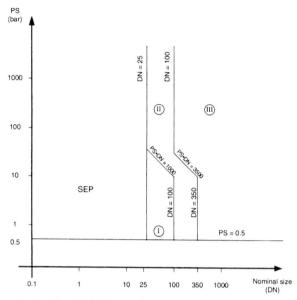

Exceptionally, piping intended for unstable gases and falling within categories I or II must be classified in category III.

Chart 6   Piping for Group 1 gases

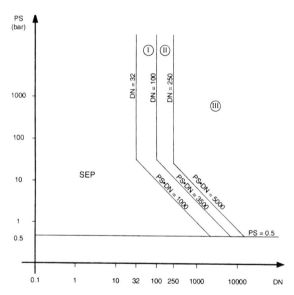

Exceptionally, all piping containing fluids at a temperature greater than 350 °C and falling within category II must be classified in category III.

Chart 7   Piping for Group 2 gases

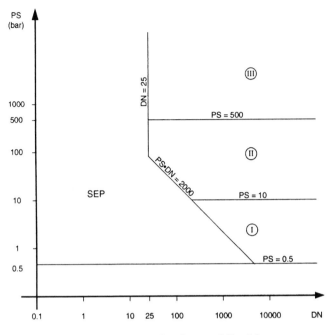

Chart 8   Piping for Group 1 liquids

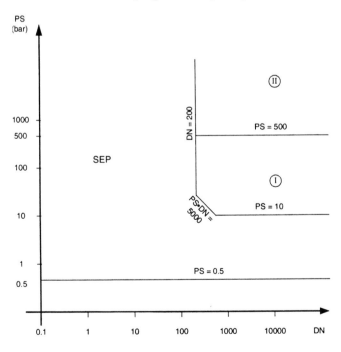

Chart 9   Piping for Group 2 liquids

Table 2.2   Applicable assessment modules

Category I module	Category II modules	Category III modules	Category IV modules
A	A1	B1 + D	B + D
	D1	B1 + F	B + F
	E1	B + E	G
		B + C1	H1
		H	

Category II equipment allows a manufacturer to choose between three different modules: A1, D1, or E1. The choice is restricted to module A1 unless the manufacturer has a quality control organization certified to ISO 9002 as required by module D1 or to ISO 9003 as required by module E1.

Category III equipment allows a choice of five modules. Note, however, that module B is intended for an EC type examination which is intended for series production of pressure equipment. The combination of B + E or B + C1 depends on the manufacturer's type of QC organization. The combinations B1 + D or B1 + F are similar but are intended for one-off fabrications. Module H is for manufacturers who have ISO 9001-approved organizations for design, manufacture, and testing.

Category IV equipment allows the choice of four sets of modules. Module G can be said to be the route normally used by EPC contractors in the past. This requires a design examination by a notified body (previously called a third party) with surveillance of fabrication through to certification.

The module requirements are summarized in Table 2.3

2.9.5   Compliance with the EHSRs

There are a number of EHSRs to be considered that involve the use of notified bodies.

Materials of construction

The PED requires the use of EU harmonized standards (EN series). If for some reason an alternative to using a harmonized standard has to be used the manufacturer can:

- Request European approval by a notified body of a particular material specification. This is not likely to succeed. Applications to approve existing common materials have been unsuccessful and it would seem that this route is for the approval of novel materials.

Table 2.3   Module requirements

Module	Design	Fabrication
A	Draw up technical file	Own QC, no third-party inspection
A1	Draw up technical file	Own QC. Final assessment monitored and random checks by notified body
B	Provide technical file for EC type examination by notified body	Verification of compliance and approval of prototype for series production by notified body
B1	Provide technical file for EC design examination by notified body	Verification of compliance and approval of one-off fabrication by notified body
C1	Applicable for series production	Notified body monitors final assessment
D	Holds EC type-examination cert. or EC design-examination cert.	Own QC to ISO 9002, QC system audit by a notified body
D1	Draw up technical file	Own QC to ISO 9002, QC system audit by a notified body
E	Holds EC type-examination cert.	Own QC to ISO 9003 for final inspection and testing. QA by notified body for final inspection and test
E1	Draw up technical file	Own QC to ISO 9003 for final inspection and testing. QC system audit by a notified body
F	Holds EC type-examination cert. or EC design-examination cert.	Product verification by a notified body examining and testing each item
G	Design examination by a notified body	QA of manufacturing through to certification by a notified body
H	Manufacturer holds ISO 9001 for full QC and QA	QC system audit by a notified body and QA of final hydro for Category III and IV vessels for Group 1 and 2 gases and Group 1 liquids for piping and steam generators
H1	Manufacturer holds ISO 9001 Design examination by a notified body for issue of a EC design-examination certificate	QC system audit by a notified body Final assessment and full QA by a notified body

- Request a particular material appraisal by a notified body for use on Category II, III, or IV equipment.

However, it should be noted that the intention is to enforce the use of harmonized standards and the use of an alternative material is only likely to be permitted in the short term where harmonized standards have not been issued.

NDT, welding, and other joining processes
Except for organizations approved to ISO 9001, a notified body or recognized third-party organization must approve welding procedures and welders for equipment in Categories II, III, and IV.

Similarly, personnel approved by a recognized third-party organization must carry out non-destructive testing (NDT) of permanent joints for equipment in Categories III and IV.

Comment
The manufacture and selling of items of pressure equipment is reasonably straightforward. In the case of assemblies it becomes more complex. The relationship between the manufacturer, contractor, and the various subcontractors or suppliers needs to be thought out and planned. A decision needs to be made as to which items are to be supplied CE stamped and which are to be issued with a declaration of conformity or incorporation. It could be that several notified bodies become involved. Some rationalization needs to be carried out to make plain the lines of responsibility for compliance and to avoid duplication and unnecessary bureaucracy. If necessary, the advice of a notified body should be sought.

2.10   Simple Pressure Vessels Directive (SPVD)

This directive is broadly similar to the other directives given in more detail above. It is limited in scope and was issued prior to the PED and is now an abnormality. This directive is under review. The products specifically have to:

- meet essential safety requirements;
- have safety clearance from an approved body;
- bear the CE mark;
- be provided with technical documentation;
- be accompanied by manufacturers' information;
- be safe.

2.11 Pressure Systems Safety Regulations 2000 (PSSR)

These regulations are concerned with the continued safe operation of pressure equipment after installation. The regulations require that the equipment be maintained in a safe condition by adherence to a written scheme of examination by a competent person. See Chapter 8.

2.12 Control of Major Accident Hazards (COMAH) Regulations 1999

These supersede and include the requirements of the Control of Industrial Major Accidents Hazards (CIMAH) Regulations of 1984. They are applicable to situations where there is a potential for a major accident as indicated by the presence of toxic or flammable substances.

The major difference between CIMAH and COMAH is the concept of a 'permissioning regime'. Once a site has been identified as a COMAH site a whole raft of processes can be imposed to ensure that the risk of a major incident is reduced. Site operations are not allowed until the competent authority has granted permission. The competent authority consists of the HSE and the Environmental Agency in partnership.

There are two tiers of requirements as judged by the HSE to pose a risk of a major incident. The factors for deciding on which tier will be applicable are complex and the HSE should be consulted in the first instance. For sites that are considered to be a top-tier risk, a risk assessment and a safety case is required. The risk assessment will have identified major hazards and their possible occurrence. The risk must also be demonstrated to be as low as reasonably practicable (ALARP). It is obvious that the requirements of COMAH should be taken into account at the planning and design stages of any potential COMAH site.

ALARP means as low as reasonably practical. The demonstration of ALARP means that the measures taken to reduce the risk can be judged to be acceptable.

The safety case will have stated the actions taken by management to minimize the risk from the hazards; for example, training, supervision, and institution of controls and procedures. The major elements of the emergency plan will stipulate the action needed to:

- raise the alarm;
- save life;
- contain the incident and prevent its escalation;
- marshal the external emergency services: police, fire brigade, etc.;
- ensure adequate training of individuals in all procedures by the staging of simulated emergencies.

2.13   Summary

Study of the above EU directives as enacted by the UK regulations will reveal a common theme. The need for compliance with the EHSRs, the requirement to retain data (a technical file) as evidence of compliance, the issue of a declaration of compliance, and the affixing of a CE mark. The idea of incorporation is established and the fact that the CE mark could indicate compliance with more than one directive. The following example will serve to illustrate these points.

2.13.1   Example of a gas turbine enclosure
A gas turbine is supplied to drive a process compressor. There are a number of issues that fall under the EHSRs of the Machinery Safety Regulations, the Dangerous Substances and Explosive Atmospheres Regulations, the Electrical Equipment Regulations, the Pressure Equipment Regulations, and the Electromagnetic Compatibility Regulations.

Machinery Safety Regulations
- The hot surfaces of the gas turbine are likely to be well over 450 °C; a safety hazard to humans.
- The noise emission exceeds the limits to avoid hearing damage.
- An enclosure is usually provided around the gas turbine. This in effect controls access as required by the EHSRs.
- The gas turbine has to be safeguarded against over-speed.
- Coupling guards are needed to protect personnel.
- The control system has to be integrated with the compressor controls and the process for safe starting and stopping with emergency shutdown (ESD).
- The enclosure has to have fire protection, usually CO_2, and fire detectors.
- The turbine enclosure has to have warning notices.
- Entry has to be controlled with safety interlocks with the fire protection system and warning alarms for evacuation. These features are needed to ensure safe entry by personnel.

- The gas turbine casings will have lifting lugs.
- To ensure correct assembly, removable guides and location dowels are fitted.
- Maintenance tools are provided for safe removal and installation of the rotor.

Dangerous Substances and Explosive Atmospheres Regulations

- The gas turbine does not normally leak gas but it could and can be classified as Zone 1. However, an ignition source is always present and for this reason it must be classified as Zone 0.
- Dilution is provided as the first means of protection. A ventilation fan provides this. This serves the dual purpose of dilution and cooling. Should this fail a 100 per cent standby fan is provided, possibly connected to an uninterruptible power supply (UPS). The fan motors must be certified as explosion proof with an EX and CE mark. In the case of the motor fan set, it could be considered to be working in a Zone 1 environment. As this is normally hot air, a Category 2 fan set could be considered.
- As a second means of protection, two out of three gas detectors are used to monitor the atmosphere inside the hood with alarms and shutdown functions.
- All the electrical equipment is explosion proof and the gas turbine can be classified to ATEX Regulations as being Category 1.

Electrical Equipment Regulations

All electrical equipment will need to comply with these regulations as well as the ATEX Regulations.

Pressure Equipment Regulations

The fuel system gas supply manifold will probably need to be Group II G for natural gas and Category 1 due to its size.

Electromagnetic Compatibility Regulations

All electrical equipment will need to comply.

Comment

It could be argued that the risk of a hot surface is continuous and the need for dilution and detection will control the risk of fire or explosion. As the enclosure is being controlled it could also be argued that the electrical systems are really working in a Zone 1 area. However, probably the notified body and the HSE will take the conservative view that everything must be Zone 0 for consistency.

The gas turbine will need to be supplied with a declaration of incorporation with the compressor and in turn with the plant pressure systems. The declaration of incorporation will need to list compliance with all the applicable regulations as listed above.

Chapter 3

Risk Assessment

3.1 Introduction

Risk assessment is a process whereby hazards can be identified so that the risks posed and their consequences can be assessed. The terms used are defined as follows.

- Hazards – are a potential to cause harm.
- Risk – is the probability that harm will arise.
- Harm:
  - to people's safety, and health and to the environment (EH&S); there is a legal duty to take action to reduce the risk;
  - arising from the malfunction of machines and processes that impact upon the economic health of an enterprise.

Very often harm has an impact on both of the areas outlined above. Even if only EH&S are involved the related consequential financial losses can be immeasurable. Large-scale disasters have destroyed businesses. In other cases the resulting loss of confidence and goodwill can affect business opportunities over a decade.

Risk assessments help to rank hazards in accordance with their likelihood of occurring and the magnitude of their consequence. Identifying the low likelihood of a large-scale disaster is as important as planning for the frequent minor incidents that simply disrupt plant operation. Both types of incident present a risk to a business and should be adequately planned for.

Two types of risk assessment are used: qualitative and quantitative. Qualitative assessments rely on the experience of a team to rank the likelihood of occurrence and the consequence that could arise from a hazard based on the situation under question. Risk ranking and the qualitative procedures used to identify the hazards applicable will be the subject of this chapter.

3.2   The information required

The first step in any risk assessment is to gather and analyse all the information on the machine that is to be reviewed. This should cover the design intent of the machine, the actual process envelope in which the machine is intended to operate, the operating controls on the machine, and the maintenance applied to the machine. Due attention must also be paid to the expected operating environment, especially its geographical location with regard to the applicable meteorological and geological data. This may affect material selection and the required output of drivers.

Reviews of operating experience and the machine's containment capability should be included with the risk assessment, which should also include a design verification report. A thorough risk assessment and design verification analysis will identify machine risks and thus actions that can be taken to minimize their impact. The information should also include a listing of the technical codes and standards used during the machine and process design phase.

3.3   Identification of hazards

Hazard assessment is the first step in risk assessment. It is the identification of all possible sources of danger to safe and reliable operation of the machinery. Some typical examples are given in Appendix B.

There are many sources from which a hazard can arise. The danger could be inherent in the process, such as handling of a poison gas. A hazard could occur from the result of a component failure. Very often it could result from operator error during transient conditions, such as starting, stopping, a change of throughput, or a change in process specification. A change of state from operation to maintenance that requires isolation and purging of toxic gas can be dangerous if proper facilities are not provided.

The Machinery Directive requires that these dangers must be identified so that design measures can be taken for their avoidance. These measures should consist of some or all of the following:

- procedures (permit systems, inspection, etc.);
- facilities (interlocks, barriers, protective systems, etc.);
- training;
- supervision and management.

3.3.1 Machine hazard assessment

Care must be taken to ensure fitness for purpose of machinery and compliance with all relevant HSE regulations and guidance notes. The Machinery Directive requires compliance with the Essential Health and Safety Requirements (EHSRs); a typical checklist for this is given in Appendix C. These requirements should be checked by a competent person with a full understanding of safety and HSE requirements in the operation and maintenance of machinery

A reliability assessment may also form a critical part of a safety assessment, particularly where the subject machine is in toxic, flammable, or explosive service. The purpose of this reliability review is to ensure that historical data are used to optimize maintenance routines and to highlight possible failure modes and their consequential effects. The results of this exercise will then be used in specifying adequate provision of spare parts and also for determining the possible need for additional safety equipment.

The importance of the machine as a part of a complete process train or unit must always be appreciated.

3.3.2 Maintenance hazards

Maintenance operations are a major risk to safety. While hazard and operability (HAZOP) studies will ensure that all the necessary facilities are in place so that maintenance can take place safely, accidents can still occur due to human error. This has to be controlled by the use of permit systems to regulate access. Procedures need to be conceived, and training and supervision provided, to ensure that maintenance work is carried out safely. In spite of this, accidents still happen and one of the areas of highest risk is when there is a change of shift. Without an adequate overlap with sufficient time for a formal handover, misunderstandings will occur.

In one case a man was still working inside the crankcase of a large engine when the second shift arrived. They thought that all work was finished and started up the engine with the man inside.

Other hazards are the result of accidents during maintenance work itself. Dropped loads and falls from heights are a common cause of accidents. It is important to provide adequate lifting facilities and lay-down areas to ensure safe working conditions. As will be discussed later, the regular testing of critical instruments is important in ensuring reliability. The locations of these instruments with adequate access facilities are critical to ensure safe and convenient maintenance. Too often this is overlooked and instruments are placed out of reach with no means of access.

3.3.3 Human interface hazards

As required by the EHSRs, machinery controls must be ergonomically designed. This means with due regard for the limitations of humans. Controls must be designed to take account of the work force characteristics; these should include, but not be limited to, the consideration of:

- *Ergonomics.* Special attention should be paid to match the required operating skills to the level of intelligence, education, motivation, and training of the work force.
- The *anthrometrics and biomechanics* of the work force to ensure that the required range of physical activity and range of vision needed are within their capacity.
- The health, fitness, and *physiology* with respect to the required speed of reaction.

Designers must take these characteristics into account in locating hand-valves and instruments. Sufficient time should be allowed for the operator to decide how to react to any abnormal situations.

> *For example the shutdown of a process may need the operation of a number of valves in a particular sequence. In an emergency there will be a risk that the operator will make a mistake. The risk can be avoided by the provision of an automatic pre-programmed emergency shutdown (ESD) system that can be actuated by a single push button.*

Plant and machinery require human intervention, however small. Tell-tale trends, signals, and alarms can give warning of incipient machine failure. Operator skill in interpreting the signs and the ability to take speedy action to advert disaster will affect the consequences of machinery failure. The location, accessibility, and ease of use of any controls will affect the reliability of its operation.

> *A common failing is to locate filling devices without ensuring that the level gauge is within sight of the operator. Likewise if the fill operation is by use of a gate valve it will be difficult to control a top up of a few litres. The fill rate needs to be slow enough to allow the operator to judge when to stop and not too slow to allow a loss of attention.*

It must also be understood that operators are inherently unreliable, just like any other control system. The consequences of operator failure must be assessed. If they are unacceptable then other systems must be introduced to reduce the risk. It is a mistake to rely only on an operator for a duty where operator error will lead to disaster.

3.3.4 Process materials hazard assessment
A vital aspect of engineering any process machine or system will be the reliable containment of the process materials being handled. Invariably these will be fluids, either gases or liquids.

A critical part of the materials hazard assessment will be the degree of risk to EH&S should containment be lost. Some fluids are a poisonous inhalation hazard, and some are flammable. Some are both, but they do not all pose the same degree of risk. The National Fire Protection Association (NFPA) publication, *Hazardous Fluids* contains a list of process materials with health, flammability, and reactivity hazard ratings. The ratings are ranked as shown in Table 3.1. The definitions, although paraphrased and simplified, provide an indication of how the ratings are ranked. It should be noted that Ratings 4, 3, and 2 correspond to the United Nations Packaging Groups I, II, and III as contained in the UN publication, *Recommendations on the Transport of Dangerous Goods.*

3.3.5 Transient and operating location hazards
Transient and location hazards will affect safety and reliability if not considered in the design of equipment. The need to lift equipment and transfer between different modes of transport during shipment and the handling requirements for erection, should be identified. There may be weight and size restrictions in shipment. The equipment may also transit through zones of sub-zero temperatures that can fracture materials and damage electronics. Equipment may need to operate in a marine environment or in areas with gale-force winds and earthquakes. Installations on ships or vehicles will experience abnormal dynamic forces. Outdoor installations may need to withstand extremes of temperature.

The local environment, in turn, may need protection from the hazards from the equipment, such as emissions and waste disposal. Where hazardous materials are involved the COMAH regulations will apply.

Table 3.1   Materials hazard rating

Rating	Possible health injury	Material flammability	Reactive release of energy
4 UN I	Death or major injury from a brief exposure	Readily burns but quickly vaporizes under ambient conditions	Possible self-detonation, explosive decomposition, or reaction at ambient conditions
3 UN II	Serious temporary or residual injury from a short exposure	Can be ignited under almost all ambient conditions	As above but needing a strong initiating source or when heated under confined conditions or materials that react explosively with water
2 UN III	Temporary incapacity or possible residual injury from intense or continuous exposure	Can only be ignited under high ambient temperature or if moderately heated	For violent chemical change needs elevated temperature and pressure. Or materials that react violently or form explosive mixtures with water
1	Exposure only causes irritation and minor residual injury	Can only ignite if preheated	Normally stable except at elevated temperatures and pressures
0	No hazard other than that of any normal combustible material	Does not burn	Remains stable even when burnt or mixed with water

3.4   Risk assessment

The simple approach of reviewing the EHSRs and taking measures to prevent the risk that they pose, as required by law, is relatively straightforward. In other situations where there is a possible risk of a disaster the law requires the risk to be as low as reasonably practicable (ALARP). A risk assessment is required.

A flow diagram illustrating the steps to be taken before arriving at a risk assessment is shown in Fig. 3.1. Acceptable risk to safety is usually judged by the chance that health and safety could be affected. Risk assessment is the act of weighing the balance between the chance of a failure occurring and the resulting consequence. This can be done using qualitative assessment based on opinion. This requires a structured approach using ranking criteria as given below.

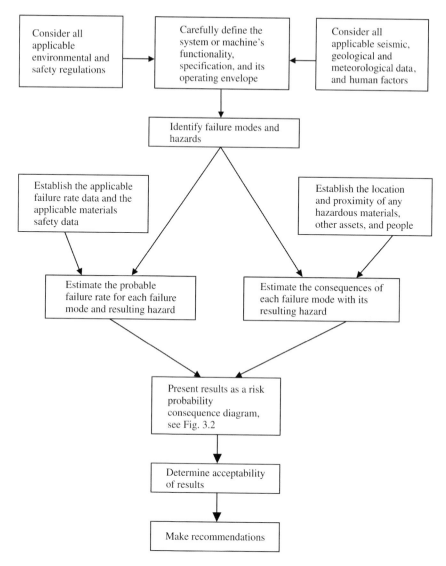

Fig. 3.1 Flow diagram for risk assessment

Risk ranking – consequence severity level

Class   *Definition (anyone or more events)*
A. Serious In-plant fatality, public fatality, extensive property damage, serious environmental damage, 2 or more days downtime.

B. High   Loss time injury, public injuries, significant property damage, environmental violation, 1–2 days downtime.

C. Medium   Minor injury, moderate property damage, minimum short-term environmental damage, 4–24 h downtime, disruption of production.

D. Low   No injuries, minor property damage, environment unaffected, less than 4 h downtime.

E. Minor   No injuries, property damage, or environmental impact, recoverable operational problem.

Risk ranking – likelihood

Class   *Frequency of occurrence*
1. Frequent   Potential to occur frequently (many times a year).
2. Probable   Potential to occur several times a year.
3. Occasional Potential to occur occasionally (once a year).
4. Moderate   Potential to occur under unusual circumstances (once or twice in a facility lifetime).
5. Unlikely   Could possibly occur, or be known to occur in the same industry, but not likely to occur over the facility lifetime.

A risk level matrix for risk assessment is given in Fig. 3.2. The result can be found by selecting the appropriate severity rating based on the possible consequences and then reading across to select the probability of occurrence. If the result is located in a shaded area

			SEVERITY				
		Frequency/h	A	B	C	D	E
LIKELIHOOD	1	$>10^{-3}$	1	2	3	4	5
	2	$10^{-3} - 10^{-4}$	2	4	6	8	10
	3	$10^{-4} - 10^{-5}$	3	6	9	12	15
	4	$10^{-5} - 10^{-6}$	4	8	12	16	20
	5	$10^{-6} - 10^{-7}$	5	10	15	20	25

Shaded area indicates no action needed

Fig. 3.2   Risk level matrix

then no management action is needed. Low numbers require more measures to be taken. The probable incident frequency is given in the table so that the perceived risk can be compared to the ALARP criteria.

3.5 Acceptable risk – ALARP

It should be noted that the principle of ALARP, as originated in the UK, is not universally accepted. For example, it is not accepted in the USA or in Germany. What is an acceptable risk is emotive and 'as low as possible' is a universally acceptable requirement. However, generally accepted criteria are:

- conforming to best practice (for fire prevention for example);
- investing £2 million to save a life.

A quantified risk of:

- 1 in a million chance of an undesired event is universally acceptable.

In the UK there is a more graduated approach in accordance with the principle of ALARP:

- 4 in a million or less is acceptable;
- 4–75 in a million needs justification;
- 75 in a million and above is unacceptable.

3.6 Failure mode and effects analysis (FMEA)

Failure mode and effects analysis is a step-by-step procedure for the systematic evaluation of the failure effects and the criticality of potential failure modes in plant and equipment. It has many different applications and can be applied at different levels of detail called indenture levels. For example, it may be used to determine the likelihood of breakdown of a gas compressor or the probability that the fire protection system fails to operate when asked to do so. At a more detailed level it could be employed to provide an evaluation of the failure mechanisms associated with a pressure sensor that could be an essential part of the fire protection system.

By analysing the failure modes of individual items the effect on machine and system operations can be identified. The need to take action and what those measures should be can be judged by the criticality of the effect identified. The criticality of the effect can be

ITEM: *Diesel Engine*			MODE: *Normal operation*						
					Failure effect				
Item	*Function*	*Failure mode*	*Failure Cause*	*Failure detection method*	*Local*	*System*	*Compensating provisions*	*Rank*	*Remarks*
Fuel pipe	Supply fuel	Fuel pipe fracture	Fatigue	Fire detection	Fuel spray	Fire	Fire protection system and auto ESD	4	Alter design to contain leak
Crank-case	Motion works	Explosive vapours	Hot bearing	Bearing temperature measurement Oil mist monitor	Explosion	Fire	Explosion doors, flame trap, auto ESD	4	Design provisions are acceptable
Motion works	Transmit power	Piston rod failure	Fracture	Inertia sensor	Unbalance forces	Self-destruct	Auto ESD (emergency shutdown)	4	Design provisions are acceptable

Fig. 3.3   Example of an FMEA for a diesel engine

ranked in accordance with the risk matrix in Fig. 3.2. An example of an FMEA table is given in Fig. 3.3. For each of the failures shown, the likelihood is moderate but the consequence is severe. The ranking is constant, as this type of analysis does not quantify how much the risk is reduced by the measures taken. The reduction of the likelihood will still be in the same band of probability. See Section 3.5 above for acceptance criteria. The criteria for acceptance are purely subjective. Judgment is based on whether or not enough measures have been taken to reduce the risk.

For example, the measures listed to reduce the risk due to a fuel pipe failure are not acceptable as there are designs available for fuel pipes fitted with outer sheaths. Any fuel leak is contained and an alarm is given. The system is marine insurance-approved for marine engines to avoid engine-room fires due to burst fuel pipes.

3.7   Hazard and operability (HAZOP) studies

These studies are used to identify the hazards that could be present in a process system such as in the design of machinery utility systems. They are intended to ensure that any hazards in operation and maintenance are avoided or that the risk from such a hazard is reduced.

HAZOP studies are usually carried out by a multi-disciplinary team in accordance with procedures given in the British Standards BS

IEC 61882:2001 Hazard and Operability Studies (HAZOP Studies), Application Guide.

HAZOP studies are a critical review of the piping and instrument diagram (P&ID) in which the machine is an element. A selection of guide words such as 'no', 'more' and 'less' are applied as deviations to a selected parameter such as pressure to study how the system will respond. This should ensure that all deviations are controlled in a safe manner. Another important guide word is 'maintenance'. This is used to ensure that all facilities needed for starting, stopping, isolation, venting, and purging are provided.

3.8 Summary

An introduction has been given to qualitative risk assessment procedures. Qualitative techniques serve a useful purpose as they provide a structured approach to risk assessment. The measures taken to reduce risk can be considered to be barriers to reduce the probability of an unwanted event. The more critical risks need more barriers to be erected to prevent their potential to do harm. There is no way to determine by how much the risk has been reduced. This can only be done by the use of quantitative techniques that will be discussed in the next chapter.

Chapter 4

Reliability Assessment

4.1 Introduction

Quantitative risk assessment is based on the principles of reliability engineering. Reliability engineering relies in turn on the gathering of statistical data on the failure frequency of common items of equipment used in process plant. Machinery is also dependent on the supply of utility services such as lubricating oil, cooling water, and control systems to operate. Failure of any of these will affect the ability of the machine to operate. Failure may also have an effect on Environment, Health and Safety (EH&S).

For example, lubricating-oil failure can lead to a hot bearing, which in turn will vaporize the lubricating oil and may cause a fire.

Data have been gathered for the items that make up systems and also for machines and items of equipment. Items do not all fail at the same time and by the use of statistical analysis it has been possible to formulate an equation to predict the probability of failure. The word probability must be emphasized. The prediction is uncertain and that is why ensuring safety is uncertain, but the risk of failure can be reduced to an acceptable level. In reliability engineering the following definitions are used [reference (1)].

- *Failure* can be defined as 'the termination or the degradation of the ability of an item to perform its required function'.
- *Reliability* can be defined as 'the probability that an item is able to perform a required function under stated conditions for a specified period of time'.
- *Probability* can be defined as 'a number in a scale from 0 to 1 which expresses the likelihood of an event'.
- *Redundancy* can be defined as 'the performance of the same function by a number of identical but independent means'.

4.2   Factors that affect the probability of failure

Reliability analysis is particularly useful during the design of a machine and can be used to establish an optimum maintenance and operation cycle for that machine. With mechanical systems, and particularly those containing rotating machinery, the prediction of reliability should take into account the considerable variance in the failure frequencies and average repair times of components. Furthermore:

- there is a need to consider and limit the effects of component degradation;
- operational, environmental, and maintenance conditions may also affect the validity of the generic failure rate data used in assessing equipment;
- on-site maintenance downtime for repair varies from one site to another;
- a subjective evaluation of additional factors, which may identify any uncertainties in the basic data, is also necessary.

These measures will ensure that predictions based on the constant failure rates adopted are sound, so that design decisions can be taken with a reasonable degree of confidence.

Take, for example, the reliability of a gas turbine; its ability to maintain rated output is affected by degradation of its combustion system, compressor, and turbine. These in turn are affected by the adopted margin between required power and the rated power, cyclic operation, frequent starting and stopping, and the site conditions such as atmospheric dirt and pollution. Furthermore, machine degradation is also affected by inlet filtration type, type of compressor cleaning system, and type of fuel used.

It is important to remember that repair time is made up of the time needed to operate a permit system, make ready for maintenance work, and the mobilization of tools and spare parts. Overall repair time will therefore vary considerably from site to site.

All the above considerations will affect reliability.

4.3   Measures to reduce the risk of failure

There are many examples of operating practices that have grown out of the desire to reduce the risk of failure. Here we see how testing and the use of parallel systems have been used to reduce failure.

Even from the early days of the industrial revolution, pressure relief valves (PRVs) were found to be unreliable. The practice was established of testing their availability by pulling on a lever to verify that the PRV was not stuck. Even today a PRV cannot be relied upon; they still need to be inspected and tested regularly on a test and calibration rig. Furthermore, other forms of pressure controls are needed to work together with the PRV to prevent over-pressure. Usually they will be set at a lower pressure to avoid conflict. The risk is reduced when more measures are installed to prevent over-pressure. Separate systems that are independent but with a common purpose are said to be parallel systems. The chances of failure are reduced because they all need to fail for the protection to fail. This illustrates the principle of redundancy with diversity.

In other cases, such as with the use of pumps, the uncertainty of knowing when they would break down and the resultant disruption was avoided by installing a spare pump. This is the principle of redundancy and is another example of a parallel system.

Control systems are notoriously unreliable. They have been the cause of machine unreliability due to unnecessary shutdowns. Maintenance engineers try to improve their reliability by regular testing and inspection. It is easy to see why they are unreliable. They usually consist of a sensing device followed by a controller that issues an appropriate output signal to some actuator that controls a mechanism. The failure of any one component will cause failure of the system. The chances of failure are additive so the total chance of failure will be greater than that of the individual elements. These are called series systems as the elements work in series.

These practices have a theoretical basis. It is possible to predict the probability of failure based on these various circumstances. Fortunately only one equation needs to be applied and the same equation can be used to evaluate all the different scenarios discussed above.

4.4 Reliability and failure probability

Quantitative risk assessment is based on basic building blocks that depend on the failure frequency of each item. The failure frequency over a standard period of time is called the failure rate.

4.4.1 Failure rate (failure frequency per unit time)

Failure rate is a measure of the frequency per unit of time that a machine or component breaks down. Failure rate data are available for most process machinery and system elements from published sources such as the OREDA *Offshore Reliability Data Handbook* (see Directory section). However, it is important to understand that these kinds of data are by their nature general and non-specific.

Failure rate data have a useful role during design when different machinery design options with different combinations of failure rate components can be compared and assessed. The probability of failure for each design combination can be compared and the comparative failure rate of each design can be estimated for consideration in arriving at an optimum design solution.

Where generic failure rate data have been used to evaluate a final outcome, these data should be adjusted to reflect the local circumstances in which the machine is to operate by multiplying the failure rate by an environmental factor. Once in operation, it is recommended that locally collected specific data be gathered from critical machinery for constant review.

In any modelling for predicting failure of a machine, failure rate is usually assumed to be constant and not affected by component degradation. After any failure, the item is assumed to be repairable and returned to service as new.

Failure rate data are usually given as frequency of failure per million hours and these units should be used for all calculations. In reliability calculations, failure rate is denoted by the symbol 'λ'.

4.4.2   Environmental and stress factors

In this example, generic failure rate can be adjusted for local conditions by the use of a factor K_1. Note that OREDA data are for offshore conditions. To adjust for onshore conditions they will need to be divided by 2; 2 is the factor adjustment for an exposed ship and 1 is for general-purpose conditions.

A stress factor, K_2, can be used to adjust for working conditions. The stress factor will account for how heavily loaded the equipment is. Equipment on part-load most of the time should be more reliable than those that are worked at the maximum rating all the time. These factors are shown in Table 4.1 [reference (2)].

4.4.3   Probability of failure

Based on statistical evidence, for repairable equipment, the probability of failure follows an exponential curve as defined by:

$$P = (1 - e^{-\lambda t})$$

Table 4.1   Environmental and stress factors

Environmental conditions	K_1	Percentage of component nominal rating	K_2
Ideal, static conditions	0.1	140	4.0
Vibration-free, controlled environment	0.5	120	2.0
General purpose, ground based	1.0	100	1.0
Ship, sheltered	1.5	80	0.6
Ship, exposed	2.0	60	0.3
Road	3.0	40	0.2
Rail	4.0	20	0.1
Air	10.0		

and when λt is $\ll 1$, i.e. very small, this becomes approximately:

$$P = \lambda t$$

These are the fundamental equations that are used to determine the probability of failure. $P(t)$ can be defined as the probability that a given item of a particular type might fail before a given time, t.

The equation can also be used to find the fractional time an item is out of service. For this purpose t must be assigned the value of the time it takes to return the item to service following a failure.

As λ is the failures per million hours, and if t is the repair time, then λt is the fraction of time it is out of service.

4.4.4   Reliability prediction

The reliability $R(t)$ is the probability that an item won't fail before time t. The probability of failure, P, plus the probability of success, i.e. reliability, R, makes the whole, therefore:

$$R + P = 1 \quad \text{so that}$$
$$R = (1 - P)$$

where the expression for P is that give in Section 4.4.3, by substitution:

$$R = [1 - (1 - e^{-\lambda t})] \quad \text{so that}$$
$$R = e^{-\lambda t}$$

4.4.5 Operational reliability

If λ is the failure rate, the number of failures per million hours, then the mean time between failures (MTBF) is $1/\lambda$, which is the period of hours that the equipment is working. It is tempting to believe that operations can be planned assuming equipment will, on average, function for this number of hours.

This assumption can be tested using the equation for reliability

$$R = e^{-\lambda t}$$

in place of t the value for MTBF can be used, $1/\lambda$

$$R = e^{-1}$$

this gives 0.37, that is 37%, or the equation for:

$$P = (1 - e^{-1}) \qquad \text{i.e. 63\% probability of failure}$$

The reliability can be improved if 0.25 of the MTBF is used for the expected operating period.

$$R = e^{-0.25} \qquad \text{this then gives 0.778 or 78\%}$$

This shows that unless the required operating period is very short compared with the MTBF, reliability cannot be achieved for critical duties. The most usual way to improve reliability is by the use of spare units.

4.4.6 Installed spare units (redundancy)

Systems with installed redundancy are known as parallel systems.

For example, a fire-water pump installed with three standby units is four items in parallel.

All four items must fail for the system to fail. In order to quantify the system failure probability, it is necessary to quantify the failure probability for each item of equipment.

If P_1, P_2, P_3, P_4 are the failure probabilities for the respective pumps, then for the system to fail P1 and P2 and P3 and P4 (pumps 1 to 4) must all fail. Using the *multiplication law of probability*, the probable failure of the system is:

$$P_s = P_1 \times P_2 \times P_3 \times P_4$$

The probability of failure of a running unit will be as given in Section 4.4.3.

Stationary standby units, however, need to be treated differently.

4.4.7   Probability of failure on demand

The failure of a standby stationary unit is unknown until it is required to be used. In order to improve its chances of being available, it has to be tested on a regular basis. This will enable any fault to be detected so that it can be repaired. The probability of failure on demand is then a function of the equipment failure rate and the test interval. If on test the equipment is found to have failed, on average it will have failed half-way through the test interval. Based on this assumption the probable failure rate of standby equipment can be found by using $0.5T$ in place of t, where T is the interval between testing (in hours). The equation for probable failure then becomes:

$$P = (1 - e^{-0.5\lambda T})$$

and when $0.5\lambda T$ is $\ll 1$, i.e. very small, this becomes approximately:

$$P = 0.5\lambda T$$

P is then the probable fractional dead time, that is the fraction of time when the unit is unable to work.

As T is the test interval selected by the designer it can be seen that the fractional dead time is reduced if the time interval between testing is reduced. The reliability of the standby unit depends on the testing schedule required. It is important this is recorded in the safety file and is adopted by maintenance/operations.

4.4.8   Common mode failure

The principle of segregation, diversity, and redundancy
Common mode failure is what can happen when you have 'all your eggs in one basket'. Having four fire-water pumps provides redundancy but if they are all electric-motor driven then the failure of electricity will cause them all to fail. If they are all located in one pump house then fire in the pump house will likewise cause all to be unavailable.

> *For example, in high-risk applications four fire-water pumps can be split into two different locations that are far apart. At each location one pump can be electric-motor driven and one can be diesel-engine driven.*

4.4.9   Fractional redundancy

One large crude oil pump with a spare can be expensive and so the use of three 50% sized units is very often used. Failure occurs when: three units fail, or when two out of three units fail.

The probability of failure is thus given by the sum of the probabilities of occurrence of each of the above two events. This can be evaluated by means of the following equation:

$$P_s = \frac{n!}{r!(n-r)!} P^r (1-P)^{(n-r)}$$

where

n is the number of items available
r is the number required
P is the probability of failure for each item

For the first event $n = 3$ and $r = 3$, therefore:

$$P_s = P^3$$

For the second event $n = 3$ and $r = 2$, therefore:

$$P_s = 3P^2(1-P)$$

The sum of the two events will be:

$$P_s = P^3 + 3P^2(1-P)$$
$$= 3P^2 - 2P^3$$

4.4.10   Control system reliability

Control systems are inherently unreliable because they depend on a number of items in the system that must all work together. The failure of any one item will result in the failure of the system.

For example, a fire prevention system consisting of fire detection, fire-water pumps, and a fire-water distribution system is three items in series.

In the event of a fire, the failure of any one item will cause the system to fail. There are three chances of failure and if P1 or P2 or P3 should fail, then the system will fail. If P_1, P_2, and P_3 are all small, i.e. $\ll 1$, then the probability of failure is given sufficiently accurately by the simple sum of the probabilities, i.e.

$$P_s = P_1 + P_2 + P_3$$

This means that the system probability of failure is very much higher than that of each of the elements.

For a small number of items in series the approximate expression for P can be used, if the values are very much less than 1.

For a larger number of items and if the full equation $P = (1 - e^{-\lambda t})$ has to be used, it will simplify the mathematics to first quantify the reliability for each item of equipment using:

$$R = e^{-\lambda t}$$

Each item must work for the system to work, so system reliability is simply:

$$R_s = R_1 \times R_2 \times R_3$$

where R_1, R_2, and R_3 are the individual equipment reliabilities.

The system failure probability will then be:

$$P_s = 1 - R_s$$

This provides the theoretical basis of why control systems are unreliable. Furthermore, many protective systems only function on demand. In order to reduce the probability of failure on demand they then require a routine schedule of testing. On a process plant with thousands of instruments it becomes a troublesome and tedious task. However, it must be understood that the safety and reliability of operations depend on ensuring that the routine testing is carried out.

4.5 Fault-tree analysis (FTA)

Fault-tree analysis can be used to identify the contributing factors and the root causes of failure. Each factor or machine part that contributes to the ultimate failure mode is assigned a probability of failure. Assembled into a tree structure, the logic tree analysis links the elements with AND/OR gates. AND gates are used to link elements where all events must occur for a fault (evaluated as a parallel system). OR gates are used to link elements where the occurrence of any one event will cause a fault (evaluated as a series system).

An analysis of the logic tree requires the use of Boolean algebra so that an expression can be found for the probability of occurrence of the top event (the undesired failure mode). Generic failure rates can be applied for each element together with estimated repair times or required inspection intervals and the probable failure rate calculated. From the results the main contributing factors can be identified and measures to reduce the failure rate can be explored.

A fault-tree diagram to evaluate the possibility of a fire is shown in Fig. 4.1. The possible risk of a fire can be found by calculating the possible failure of each event starting from the one at the bottom. The first event to calculate will be the probability of the electric water pump failing. This will be when both the No. 1 AND No. 2 electric fire-water pumps fail.

4.6 Factors affecting reliability assessment

The operating regime, environmental conditions, and maintenance strategy, as already mentioned, are the factors that need particular attention in mechanical system reliability assessments. Failure mode and effects analysis, in association with fault trees and a subjective analysis of equipment maintainability, should ensure that the potentially critical failure modes associated with material degradation, such as with seals and packing, are identified at the design stage.

Appropriate condition monitoring systems should be specified to ensure that these potentially critical failures are revealed before components deteriorate significantly. Maintenance actions can then be planned for the next time the plant requires a shutdown.

Planned maintenance can be used based on predicted availability if degradation-type failures are anticipated. The eventual corrective action can then be pre-planned. This will then be the best that can be achieved without introducing additional redundancy or other design changes. However, the failure tolerance of alternative systems, such as the use of glandless pumps, can in some cases be significantly higher and some loss of process efficiency may be a worthwhile price to pay for a system with a higher availability potential.

4.7 Reliability in design

Process plant and machinery design evolve through distinct phases. The initial stages are the most critical to a reliable design.

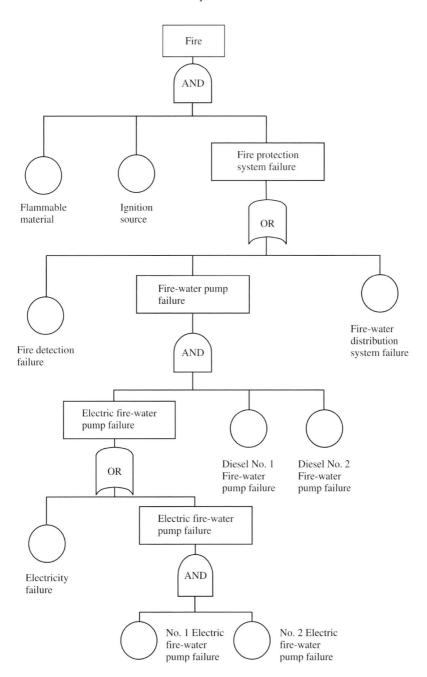

Fig. 4.1   Fault tree for a fire

Initial studies

These establish the need for a plant and its purpose. At this stage it is also important to consider its economic future. Subsequent changes in its operational use could prove costly if they are not considered at the design stage. The conceptual design of the process machinery and the plant are intertwined. Process stream size could depend on available machine capacity. The need to accommodate incremental increases in throughput will affect machinery arrangements.

Conceptual studies

In order to match the process operating requirements, the type, size, and arrangements of machines will need to be studied. At this stage, due to process needs, machines of smaller or larger sizes than normal may need consideration. The design hierarchy of increasing risk will be:

- well proven in similar service over many decades;
- only a few in operation with limited running experience;
- interpolation of an existing design;
- extrapolation of an existing design;
- use of new technology.

Design specification

The reliability of any design selected will depend on its operability and maintainability. Components and sub-assemblies should be categorized as:

- wearing parts;
- aging parts;
- parts that are subject to mechanical failure;
- parts expected to last the lifetime of the machine.

For each item there will be an expected life, i.e. failure rate or MTBF. The specification must prescribe the required actions to ensure that the expected failure rates of the different components will add up to that needed to provide the required reliability of the whole machine. More research and development testing could be needed on selected items.

One area of concern will be the reliability of instruments and their maintainability and operability.

Another area of concern will be the need for condition monitoring to ensure the maximum operating time before maintenance.

It will be necessary to ensure that the communication gap between the purchaser's needs and the supplier's understanding is reduced to a minimum. This should be audited from the management level down to the shop-floor supervision. A co-ordination meeting that explores the understanding of each clause in the specification will do this. The design process can then be audited and controlled by an approval process for critical drawings and procedures ahead of manufacture.

Manufacturing QC/QA

Even well-proven designs will have elements of risk that will need to be identified. All manufacturing processes should be reviewed. Defects in fabrication and manufacture could be a high risk due to new methods or the use of a new subcontractor. The possible failure of QC and QA is a risk that can be reduced by imposing more stringent procedures and additional third-party inspection.

Factory acceptance test

Close supervision of final assembly and testing is the final safeguard to ensure that a reliable machine has been designed and made. As far as possible everything that has been specified should be measured by some acceptance criteria.

A machine enclosure was specified to be protected from driving wind and rain. A simple hose test on final assembly revealed its ineffectiveness.

Installation, start-up, and operation

The final test will be in use. Reliable operation may still need some development due to unforeseen and unusual circumstances. An experienced engineering team will need to be available to trouble-shoot and decide on the most expedient means of resolving problems. At this stage accurate logging of events and recording of outcomes will be necessary to ensure progress.

It is important to realize that success in ensuring a reliable design will depend on how much work is done before issuing an enquiry and then studying proposals before order placement. Any attempt to roll much of this work on to the supplier with the purchase order can only lead to extra cost and delay, if not an unsatisfactory outcome.

4.8   Reliability specification

It is recommended that for all large or complex process plant machinery, a reliability specification is included in the purchasing specification if the likelihood and consequences of failure will have a major impact. Model clauses, which may represent a reliability specification, are detailed later. The object of such a specification is to ensure that consideration is given to reliability and availability in order that:

(a) a minimum requirement is achieved;
(b) opportunities to optimize the design for reliability and avail-
    ability are not overlooked.

Wherever possible, reliability targets should be set:

(a) to match, or better, those of similar machinery;
(b) to meet purchaser's needs in relation to a combination of risk,
    cost, and benefit.

It is recognized that it may not always be possible to set formal targets, and in this case the purchasing specifications should still require reliability/availability assessments to be executed and the machinery design reviewed for possible improvements. Cost–benefit analysis should then be applied to enable sound judgments to be made.

As part of the purchasing cycle, reliability specifications should be included in the enquiry documents to ensure that the selection considers reliability/availability. The potential supplier, even if they do not have information to hand on reliability, should be required to state in their tender how they intend to address this topic in the event of an order. It may require additional payment that should be identified and agreed before order placement.

It is stressed that such assessments should not cause major or insurmountable problems to any manufacturer or supplier. By referring to BS 5760, published works, or consultants, a manufacturer or supplier should be able to use these techniques to facilitate conclusion of a satisfactory order. Historical data on similar machines may at first appear to be scarce, but data banks do exist in the public domain and obviously machines are constructed of many simple parts for which data are available.

4.9 Model safety and reliability specification

It is recommended that the following model clauses (as published by the Engineering Equipment and Material Users Association, EEMUA) should be included in purchasing specifications for all major items or packages of capital equipment, particularly in any case when the extent of supply includes anything that could be considered a 'system' (for example, the lubrication system supplied with a rotating machine).

By paying specific attention to reliability issues that affect health and safety, this in effect reinforces the requirements of the Machinery Directive.

(1) The vendor shall demonstrate that they have considered the reliability/availability aspects of the package they are proposing to supply in a disciplined and structured manner. The object of this consideration shall be:
   (a) to identify critical failure modes, especially those that have an impact on health and safety;
   (b) to estimate the overall reliability/availability of the total package in relation to safety issues and the duty or duties which it is intended to perform;
   (c) to identify those components of the package that make a major contribution to unreliability, unavailability, and risk to health and safety;
   (d) to provide confidence that major potential weaknesses in the package have been estimated.
(2) An acceptable procedure which would satisfy the above requirements would consist of:
   (a) construction of a reliability block diagram where each 'block' represents an item, component, or sub-system for which meaningful reliability data can be provided;
   (b) identification of the potential failure modes of the 'blocks' and the consequences of such failures on the performance of the package;
   (c) by using a fault tree, or equivalent alternative method, computation of the overall unreliability/unavailability of the package and identification of significant contributors;
   (d) proposal of modifications for consideration by the purchaser to improve the design.

(3) The vendor shall declare the data used, the sources thereof, and any assumptions made. The purchaser may wish to review this data and discuss the applicability to the particular equipment. The vendor may be required to substantiate any data that are not within the public domain.

(4) In making assessments of availability, the vendor will need to estimate repair times. For the purpose of making such estimates, the vendor may assume that the spares holding is in accordance with his submitted recommendations unless otherwise stated. He may also neglect any delay caused by factors not directly related to the package in question, unless a basis for estimating the extent of such delays is provided by the purchaser. In the case of failure modes that make a significant contribution to the overall unavailability the vendor may be required to demonstrate that their assumed repair times can be achieved.

(5) The overall availability/reliability of the whole package, as predicted by the analysis, shall not be less than the value stated in the enquiry or order, or such other value as may be agreed.

(6) The vendor's warranty for the package shall not end until such time that the reliability/availability achieved in an agreed trial period has been demonstrated to be not less than that predicted, or some other value as may be agreed. The length of the trial period shall not be less than 10 times the predicted overall mean time between failures, or some other period as may be agreed, but in no case less than 6 months.

Comments on the above specification

Difficulties will usually be experienced in obtaining guarantees and/or warranties with respect to machinery reliability. Detailed negotiations should be expected with particular reference to the issues as given in Section 4.3 above. In some cases, to overcome these problems and in order to obtain a guarantee, it may be possible to place a contract for maintenance, or in some cases total operations and maintenance, with the vendor. As a minimum, however, vendors can be required to guarantee and demonstrate repair times as suggested in paragraph (4) of the above specification.

4.10 Summary

An introduction has been provided on reliability assessment in sufficient detail for an understanding of the basic principles. It has been shown that reliability is achieved by the reduction of risk. Risk has to be identified and controlled by taking steps for its reduction.

An outline has been given of the application of reliability assessment to the specification of machinery and how it relates to the design process. It should be apparent that this work requires specialist knowledge and experience with a good understanding of the design and characteristics of all types of process machinery. This requires the skills of a multi-disciplinary team, as will be shown in later chapters.

An example of reliability analysis follows to illustrate the application of the equations given in this chapter.

4.11   Reliability analysis – an illustration

4.11.1   The scenario
Utility air is to be provided for a process plant to be designed for 3 years continuous operation. Any loss of air supply is considered to cause a serious loss of production and a reliable system is required. The data for a centrifugal air compressor package are as follows:

   Failure rate is 300 failures per million hours
   One year is 8760 hours
   Failures are assumed to be minor in nature and can be repaired
      immediately
   The time to repair is one shift of 8 hours
   The length of reliable operation required is 25 000 hours
   Test interval, T is 722.5 hours (a one-month interval being a
      convenient option) for the standby machinery

Installing one machine
The fractional downtime is $P = (1 - e^{-\lambda t})$ or approximately $P = \lambda t$.

   $$P_a = 300 \times 8 \times 10^{-6} = 0.0024 \text{ fractional downtime}$$

The probable downtime for 25 000 hours is:

   $$0.0024 \times 25\,000 = 60.2 \text{ hours}$$

Its reliability, $R = 1 - 0.0024$.

   $$R = 0.9976, \qquad \text{i.e. } 99.76 \text{ per cent}$$

For a critical air supply this is not acceptable. For the period of

25 000 hours a downtime of 60 hours will represent a probable seven to eight failures.

Installing one machine with one standby
To improve reliability a spare compressor is needed; assuming the second compressor is on standby and is tested once every 722.5 hours, the probable failure rate is therefore:

$$P_b = 0.5\, \lambda T$$

which is

$$0.5 \times 300 \times 722.5 \times 10^{-6} = 0.108\,375$$

This means a 10.83 per cent probability of failure, i.e. 89.97 per cent reliable.

The failure rate for the two machines, as they must both fail for no air supply, will be:

$$P_s = P_a \times P_b = (0.0024)(0.108\,375)$$
$$= 0.000\,26 \text{ fractional downtime}$$

System probable downtime for the required mission time is:

$$0.000\,26 \times 25\,000 = 6.5 \text{ hours}$$

Reliability:

$$R = 1 - P = (1 - 0.000\,26)$$
$$R = 0.999\,74, \qquad \text{i.e. } 99.974 \text{ per cent}$$

Better but not good enough. It is highly probable that at least one total failure will occur during 25 000 hours of operation.

Installing one machine with a running spare
Both machines will have the same failure rate and the fractional downtime will be:

$$\text{System } P_s = P_a \times P_b \qquad \text{(one machine running with a spinning reserve)}$$

$$\text{System } P_s = (0.0024)(0.0024)$$
$$= 5.76 \times 10^{-6}$$

which is the system fractional downtime.

The probable downtime for the required mission time is the system fractional downtime multiplied by the required duration, which is:

$$(5.76 \times 10^{-6}) \times 25\,000 = 0.144 \text{ hours}$$

The reliability of the system is:

$$R = 1 - 5.76 \times 10^{-6} = 0.999\,994\,24 \text{ or } 99.999\,424 \text{ per cent}$$

To ensure a reliable air supply for the required operating period of 25 000 hours requires one machine and a running spare to be installed.

The use of two 50 per cent machines with a running spare
As previously discussed it may be cheaper to install three 50 per cent-sized machines instead of full-sized machines. This means that two out of three need to be running. This can be evaluated by means of the following equation:

$$P_s = \frac{n!}{r!(n-r)!} P^r (1 - P)^{(n-r)}$$

where

n is the number of items available $= 3$

r is the number required $= 3$ and 2

P is the probability of failure for each item, which is $P_a = 0.0024$ as before

The first failure mode is three to fail when $r = 3$:

$$P_1 = \frac{3!}{3!(3-3)!} P^3 (1 - P)^{(3-3)}$$
$$= P^3$$

The second failure mode is two to fail when $r = 2$:

$$P_2 = \frac{3!}{2!(3-2)!} P^2 (1 - P)^{(3-2)}$$
$$= 3P^2 (1 - P)$$

The probability of system failure is the sum of the two modes of failure:

$$P_s = P^3 + 3P^2(1 - P) = 3P^2 - 2P^3$$

and as $P = 0.0024$ is very small, the last term can be disregarded:

$$P_s = 3P^2 = 3(0.0024)^2$$
$$= 17.28 \times 10^{-6}$$

which is the system fractional downtime.

The probable downtime for the required mission time is:

$$17.28 \times 25\,000 \times 10^{-6} = 0.432 \text{ hours}$$

The reliability of the system is:

$$R = 1 - 17.28 \times 10^{-6} = 0.999\,98 \text{ or } 99.998 \text{ per cent}$$

This result is based on all three machines running so that there is a spinning reserve.

The use of two 50 per cent machines with a running spare and a standby
This can be considered to be a two out of three system in parallel with a standby.

As before, if the probability of failure of the two out of three system is:

$$P_s = 3P^2 = 0.108\,375$$

the probability of failure for one standby machine as before will be:

$$P_b = 0.5\lambda T = 0.108\,375$$

For the combined system to fail, the two out of three system AND the standby must fail. The probability of failure for the combined system will then be:

$$P_{\text{combined}} = P_s \times P_b$$
$$= (17.28 \times 10^{-6})(0.108\,375)$$
$$= 1.872\,72 \times 10^{-6}$$

The system reliability is then:

$$R = 1 - 1.872\,72 \times 10^{-6}$$
$$= 0.999\,998$$
$$= 99.9998 \text{ per cent}$$

The above results show that for the period required, any machine is likely to fail. In the case of a spare machine on standby, the probable time that it is unavailable is half the time between the test periods. By inspection, even if the test period for the standby machine were to be once a week the reliability will not be acceptable. This explains why the use of a running spare (sometimes known as a spinning reserve) is adopted for critical services. A two out of three system with a running spare could be used if the risk is acceptable.

In some cases this may not be acceptable and the addition of a standby is adopted. It is interesting to note that the use of 50 per cent machines with a running standby and a spare has the same installed capacity as the use of a 100 per cent machine with a running spare. This arrangement, however, gives a slightly improved reliability at the expense of increased maintenance.

The risk in all cases can be reduced by the installation of air storage vessels. This will then enable the plant to operate for a short period without an air supply. The causes of compressor unreliability could be reviewed to identify areas for improvement. Instrumentation may be a significant cause of failure and can be improved as discussed in the next chapter.

To summarize, the results are shown in Table 4.2 below.

Table 4.2   Reliability analysis results

System installed	Reliability (%)
One compressor	99.76
One compressor with one standby	99.97
One compressor with a running spare	99.9994
Two out of three with a running spare	99.998
Two out of three with a running spare and standby	99.9998

To ensure reliable operation it is necessary to get as close to 100 per cent as possible. It is interesting to note that by the ALARP criteria the last three arrangements would be classified as a tolerable risk.

4.12   References

(1)   **Andrews, J. D.** and **Moss, T. R.** (2002) *Reliability and Risk Assessment*. Professional Engineering Publishing, ISBN 1 86058 290 7.

(2)   **Davidson, J.** and **Hunsley, C.** (Eds) (1994) *The Reliability of Mechanical Systems*, second edition. Professional Engineering Publishing, ISBN 0 85298 881 8.

Chapter 5

Control of Risks

5.1 Introduction

The combined effect of the CDM Regulations and Machinery Directive is the requirement to ensure that the needs of safety are taken into account from project conception through design, constructions and into service. In the procurement of any machinery that may be required, each of the parties concerned have specific responsibilities to fulfil.

A procedure is proposed for the control of hazards by a system of machine classification during design and verification that the design is safe before operation.

5.2 Responsibilities

5.2.1 The client

The concept of any project originates from the client. The client specifies the duty and rate of output from the project, its location, and the way in which it will be operated. This results in the specification of the required machinery. The client also has the ultimate authority on deciding the machinery supplier to be selected.

When the machinery is accepted into service the client then has the responsibility for the operating and maintenance policy. The client must ensure operation within design limits, control all risks as identified in the safety file, and carry out all necessary maintenance.

5.2.2 The designer

The designer converts a concept into a detailed design. The designer determines the detailed specification of the machine, and all the operating conditions to which it will be exposed, both transient and steady state. The designer also determines the type of machine required and may select suitable machinery suppliers for consideration by the client.

5.2.3 Project responsibilities

It is the responsibility of the design project manager to ensure that the appropriate machine system classification has been completed in accordance with the Machinery Directive.

The design project manager, in consultation with the client, is also responsible for the appointment of a safety planning supervisor in accordance with the CDM Regulations.

5.2.4 Safety planning supervisor

The safety planning supervisor is responsible for the health and safety plan and preparation of the safety file in accordance with the CDM Regulations. Typical contents will include:

(1) The safety team organization and assignment of duties.
(2) Procedures established to identify hazardous materials of construction and operation.
(3) Lists of required safety drawings to include:
    (a) hazardous area drawings;
    (b) noise protection zones;
    (c) a list of hazardous substances and their location together with the first aid services provided such as safety showers and eye baths, etc.;
    (d) location of other hazards such as high voltage, heat radiation, dangerous entry, etc. and safeguards provided to prevent entry;
    (e) rescue equipment and their location;
    (f) plot plans showing escape routes and means of escape;
    (g) facilities to signal evacuation and their location;
    (h) provision and location of emergency services and the access routes into the plant.
(4) Scheduled meetings to:
    (a) HAZOP one-line (P&ID) drawings;
    (b) HAZOP one-line electrical drawings;
    (c) make provisions for maintenance access and lifting facilities;
    (d) review constructability and safety with design and construction;
    (e) co-ordinate construction planning with engineering output, to avoid work congestion on site and to ensure the safe lifting of large objects.
(5) Classification and identification of critical machinery and processes and the nomination of equipment for risk assessment.

(6) Making adequate provisions for safety management and training during construction.
(7) The maintenance of records and the preparation of a safety file for handover to the client.

5.2.5 Machine design responsibilities

For critical machine systems it is the responsibility of the design engineer to oversee the integration of the machine system into the process. This includes:

(a) the technical specification of the machine system;
(b) choosing the supplier and specifying any inspection requirements;
(c) identification of hazards and failure modes;
(d) agreeing with the supplier on any protective systems and operational or maintenance requirements, to ensure safe operation and to meet legal requirements;
(e) obtaining design verification from a specialist engineer;
(f) specifying commissioning tests to ensure the safe functioning of the system.

5.2.6 Safety or loss-prevention engineer

The safety engineer is responsible for carrying out the safety assessment and reviewing all identified hazards and failure modes in accordance with the health and safety plan as prepared by the safety planning supervisor.

The safety engineer monitors the safety plan, chairs all HAZOP meetings and safety audits, and prescribes steps to ameliorate risks to health and safety.

For small projects this work can be amalgamated with that of the safety planning supervisor. For large projects there may be a team of safety engineers.

5.2.7 The machine manufacturer

It is the responsibility of the manufacturer or his representative to design the machine system to meet the specified duty points. They should provide a complete statement of design limits that will ensure machine safety relative to all identified failure modes.

The manufacturer shall be required to affix a CE mark to the machine and issue a declaration of conformity in accordance with EEC Directive 89/392/EEC (UK Statutory Instrument SI 1073/94).

It should be noted that, if loss of machine containment is to be avoided, provisions additional to the equipment package might be needed, such as protective devices to alarm, trip, or isolate. In such cases the plant designer and manufacturer will need to determine who holds the master technical file. This establishes responsibility for carrying out the work of incorporation and applying the CE mark.

5.2.8 The construction contractor

It is the responsibility of the site construction manager to ensure that the safety plan has been studied and the required precautions, procedures, and training of construction workers are carried out. Construction input will also be required during the design phase to ensure that the requirements of the safety plan can be implemented. Refer to the CDM Regulations for the responsibilities and requirements.

5.2.9 The construction/commissioning engineer

The commissioning engineer shall ensure that, prior to commissioning, the equipment and its installation are checked for compliance with the design intent.

It will be important to ensure that tests, as specified by the machine designer, are carried out to prove the correct functioning of the system before hazardous materials are introduced to the plant.

5.2.10 The site (plant) manager

This in effect is the client or the client's nominee. It is his responsibility to verify that the plant and its machinery are fit for purpose before accepting them into service. On handover he receives the safety file and any technical files for safekeeping and review. He then becomes responsible for the safe operation and maintenance of the plant and its machinery. See Section 5.2.1.

5.3 Classification of machines

In this context, classification is a procedure for ranking machines with regard to risk. The potential hazards of the machines are considered and the machines classified as either 'critical' or 'non-critical'. This judgement is based on assessment of the hazard presented by the process fluid and on the potential consequence of its release. Consideration must also be given to the significance of process consequences following mal-operation of the process or failure of the machine, and the potential damage due to mechanical failure given the nature of the machine assessed. The classification diagram shown in Fig. 5.1 illustrates this.

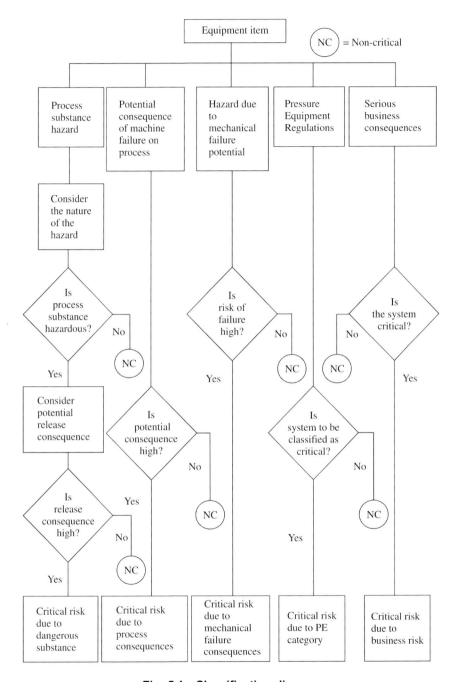

Fig. 5.1   Classification diagram

5.3.1 Significance of classification

The classification process described achieves two objectives, firstly the prioritization of the machines as to overall hazard, and secondly the establishment of criteria against which the verification needs to be conducted.

The detailed procedure for examining a machine is built around a series of reviews. In order to determine the focus of these reviews, i.e. to decide which sort of review to carry out, it is vital to understand and agree the basis for the criticality classification. This may require clarification of the original intention of an installation, where this is not apparent from the information that has been presented.

Example

The classification of a pump as critical for reasons of process consequence required the verification engineer to have a clear understanding of what these 'process consequences' were. In this particular case the machine appeared to be adequately matched for the duty. However, the process intent was such that a failure of the pump would have unacceptable safety and reliability consequences. Though the pump could be expected to be reliable, pump failure could not be ruled out.

The design had not included any particular monitoring of process or machine parameters to give prediction of such failure, and had not included the installation of a spare or of a system to provide cover in the event of failure of the on-line unit. When the verification engineer communicated this information to the project design team, it was appreciated that the plant design had not implemented the original intent of the process engineer.

5.4 Machine verification

The verification process looks at the machine and the process in which it operates. It therefore provides an independent review prior to commissioning of key areas of the production process. This systematic review identifies points of concern where the proposed equipment does not meet the original requirement.

The classification of machines takes place alongside other design reviews at the start of the design process. The verification study provides a detailed check at the completion of design, manufacturing, and installation of the equipment. It can ensure that the original intent for the equipment is met safely.

The verification review forces the design process to be complete prior to commissioning, and ensures that all design needs have been addressed.

Detail is important to successful machine operation. For example, incorrect package units (fully assembled and complete) may be supplied because materials were wrongly specified on the purchase order or on the manufacturer's standard. This will only be detected once the machine dossier is submitted. It should be detected by the design engineer, but this is not always the case because the engineer at the end of a project may not be the same one as at the start.

Consistency of intent is difficult to achieve where the equipment called for is part of a front-end engineering package and is designed by the manufacturer from a design contractor's specification. The verification engineer needs to review the whole concept of the design, bringing the key safety points into the limelight and throwing deviations from the original intent into relief.

Verification methodology
The verification analysis of the information provided must be structured so as to facilitate a logical and complete assessment. The structured information that will then be available can be reviewed as a whole to provide a consistent picture of the capability of the machine system.

5.4.1 Machine purpose and specification
This information is required to determine the suitability of the machine for the intended task. This would ideally come from three sources.

(1) *The process data sheet.* Outlines the scope of the process design.
(2) *The mechanical data sheet.* Where the process design is translated into machine requirements incorporating the safety and other best-practice points for machine reliability relevant to the particular installation.
(3) *The machine supplier's data.* The translation of the machine requirements by the supplier must be sufficiently detailed to show that the machine will be capable of meeting the full range of conditions (including start-up and shutdown) required by the process.

In the case of a new installation all this information must be available and should be supported in particular instances by design reviews of the installation, such as material-of-construction reviews, relief-and-blow-down reviews, etc. This is illustrated by the following example.

In the case of the design review of an ammonia screw compressor, it was found that the design logic of the pressure relief of the compression system, in accordance with normal procedure, was built around the capacity of the machine. However, screw compressors are designed with an internal compression ratio that is unaffected by any discharge pressure relief. It was found that high suction pressures could occur at start-up, which could result in internal pressures exceeding the casing design pressure. Additional operating reviews and manufacturer reviews were needed before the safety of the installation could be demonstrated.

5.4.2 Machine operation and protection

As with the machine purpose and specification exercise above, the information on the machine operation and protection can be found in several areas.

(1) *The engineering line diagram.* An essential document because it fixes the position of the machine in a frame of reference for the process. Examination of this information has much in common with the techniques used for hazard studies, where the normal and extreme conditions must be considered against different operating situations. Though the formal hazard study reports will be considered later in the review process (if available) the view of the line diagram from a 'machine's point of view' can give a different aspect from that which is derived from the formal study, which tends to cover process streams rather than machine effects.

(2) *Instrument schedules.* These documents provide a wealth of information in terms of confirmation that the protection identified on the line diagram is in place and that the actual limiting values associated with a protective device are correct. The reason for the device is sometimes obscure and it may be necessary to add notes on its function.

Cases have occurred where the protective device was available and in use, however examination of the instrument maintenance schedule showed it to be untested. The reason for the protective

*device was not understood and the need for it to be regularly tested
to ensure its reliability was therefore not recognized.*

(3) *Hazard and operability reviews.* These have been produced for
projects over the last 20 years, and can be extremely helpful to the
verification engineer. He will need to give careful consideration,
however, to operating experience, and to plant modifications and
changes in the intent that have occurred since the time of the
original reviews.

(4) *Operating instructions.* A full understanding of a process may
only be possible when the operating intent and practices are
known. Matters such as start-up and shutdown conditions,
remote start-up, parallel operation, or changes in feed material,
are covered in these documents and may not be apparent from
other information.

*In the case of a reciprocating compressor the operating instruc-
tions showed that, while normally protected from overload by a
low-suction-pressure alarm, it was necessary to run with this alarm
disabled during a regeneration cycle. Once recognized, the danger
of not having the alarm enabled during normal operation became
apparent. Changes to the instrumentation and operating practice
were made to ensure that the correct protection was in place for all
modes of operation.*

5.4.3 Machine maintenance and operation

Operating experience
Information on the history of a machine must form part of any
assessment. With new installations, information must be available
from the manufacturer identifying the possibilities for serious failure
and the measures that have been adopted to avoid them.

In the case of existing equipment where the machines are complex
or where there is limited experience of them within the organization,
information needs to be sought from the manufacturer on the
possibilities for serious failures, i.e. those which could threaten the
integrity of the equipment or which could lead to unacceptable
situations.

The operating history of the equipment should highlight the
occurrence of serious failures and all classified incidents must be
reported.

The information gained here will give an indication as to how well suited the machine is to its duty. Where satisfactory experience exists and is documented, this may obviate the need on existing equipment for checks on piping loads and inspections.

Operating instructions

The scope of the operating instructions covers items essential to the protection and the reliable operation of the machine. However, from observation, experienced operators can usually develop ways of carrying out simple checks and inspections of a machine which are more effective than those given in operating instructions. This experience is invaluable and should be recorded and not lost in the event of any restructuring of the operating teams. In many ways these checks can prove superior to the use of sophisticated instrumentation.

Within the organization this knowledge base may be strengthened by the integration of operating and maintenance teams. This gives an opportunity for improved surveillance gained from a better level of technical knowledge of the machine.

Maintenance instructions

Planned maintenance instructions must be based on a considered maintenance policy in which all required checks are fully identified. Reliability analysis of the machine and its systems will identify the critical parts of the machine and will aid the determination of the frequency of maintenance needed if the required level of reliability is to be sustained. Given that the required level of safety is attained, maintenance policies must also be selectively applied for optimum cost effectiveness.

For some items reliability analysis will show that the application of a given maintenance action should be usage-based (e.g. after processing a given volume of material) or time-based (usually appropriate for safety devices). In other cases, the need, or otherwise, for maintenance can be predicted by regularly monitoring appropriate parameters of operation (e.g. achieved pressure) or condition (e.g. bearing vibration amplitude), thus avoiding premature shutdown for inspection or overhaul. Where failure is not critical (the machine has a standby, for example) maintenance can be on a breakdown basis.

Whatever the maintenance policy, the instructions must be clear about the inspections required at the time of any overhaul.

In all cases good maintenance practice, carried out by appropriately trained personnel, is assumed. Specific checks are identified for critical maintenance activities, where problems have been found to be generic to certain types of machines. Examples of this are the checks on the throttle bush clearance for a hydrocarbon duty pump, checks on foundation bolting, and the assembly of the piston rod to crosshead connection on a reciprocating compressor.

Manufacturer's instructions will be taken into account during the verification and, where the operation described is key to the machine integrity, rigid adherence to particular safety elements will be specified in the verification report. There is scope for a degree of interpretation of such instructions (within the bounds of good maintenance practice) but all repairs and modifications to a machine will need to be authorized.

5.4.4 Hazard review

This is a review with a particular focus on aspects of the machine design; in some cases this will require multiple reviews, as there are several specifics to consider. The reviews are generally based on the engineering guides or specifications produced by the operating company. Though in general these can be a simple 'yes or no' review, judgement is required when considering new-technology equipment or assessing mitigating circumstances for equipment installed prior to the review.

5.4.5 Containment review

To facilitate this review a series of structured checks on the machine design should be devised. This should take account of the majority of containment situations found on common machines so as to be adequate for most of them. However, these reviews may not be enough; experience shows that the verification engineer must then make a judgement on the overall design based on experience and analysis of the machine. This should be done at the end of the design process because the items of information will only come together when the full design of the machine is understood, its history is known, and the protective system is defined.

A modification of an agitator was under review. The modification only involved changes around the bearing support and, as far as it went, was satisfactory. However, the verification engineer continued to have concerns about the equipment because the restraint of the agitator depended on only three bolts and operating history showed that there was evidence of the shaft moving in the coupling, designed to be a sliding fit, and failure of such restraining bolts had been found on other installations.

The consequences of failure would have been severe; there was a possibility that the shaft might then drop through the seal, resulting in an unconstrained release of flammable vapour from a major vessel. Redesign of the modification provided both a way of addressing the original intent, which was to improve bearing life, and of eliminating the threat to containment.

Chapter 6

Operations and Maintenance

6.1 Introduction

Hazards to a plant designer are perceived, but to a plant operator they become realities.

The plant operator must first understand what was in the mind of the designer. He must then relate it to the reality of the plant and the resources available to him. The operating situation may be different to that envisaged by the designer. The resources in maintenance, in people and materials and the actual skill of the operators, will need to be assessed against the tasks that need to be carried out.

The risks to safety have to be managed throughout the life of the plant. The probability of failure increases with time due to the complacency of operators and due to plant deterioration. The measures needed to ensure safety must be audited on a regular basis so that methods and procedures are kept up to date. Improvements to plant and machinery must be carried out with proper integration into the existing safety systems.

6.2 Handover

Before completion, and prior to operation, the legal documents for the plant must be formally handed over to the client so that staff training and maintenance planning can be based on their contents. Meetings must be held with all involved parties so that inherent risks and safety provisions are fully understood. The legal documents as prescribed are detailed below.

The safety plan
This is the document listing all the actions that were taken during the design and construction phase of the project in accordance with the CDM Regulations. The plan will need to be extended to include all

the actions needed to ensure safe operation of the plant so as to safeguard the health and safety of the workers and the general public.

The safety file
This contains a compilation of the design features provided for the health and safety of the plant workers resulting from the execution of the safety plan during design and construction, as required by the CDM Regulations. This is, in effect, all of the work as outlined in the risk assessment, see Chapter 3.

The technical file
The Machinery Directive (see Chapter 2, The Law) requires the technical file to be held by the supplier. The regulations do not require it to be issued to the client.

For process plant machinery, however, it is imperative that the client is made aware of the contents of the technical file and it should be specified to be supplied as part of the purchase order.

Process machinery supplied in accordance with API specifications usually has to include operating and maintenance manuals and QA documentation. These normally contain:

- operating instructions;
- verification reports;
- maintenance instructions and inspections;
- manufacturing records and data sheets;
- design limits of the machine;
- identification of hazards and failure modes;
- safeguards provided to ameliorate the effects of hazards and failures on health and safety;
- schedule of protective systems;
- availability and reliability data.

The required content of the technical file is similar to the above. For the process industries, it would seem that the technical file and the technical manuals should be made one and the same, to be held by the client as well as by the vendor.

6.3 Reliability-centred maintenance (RCM)

Any machine, however well designed, must be maintained to ensure reliable safe operation. CDM Regulations require maintenance to be considered from the earliest stages of a project. Requirements and targets must be established at the start of a project when budgets and resources are being allocated. This ensures that maintenance facilities will be provided for in the plant design.

RCM is an ideal tool which, if applied at the design stage, will lead to optimizing planned maintenance and managing the risk to safety in operation.

In the reliability assessment of the machine the various failure modes will have been identified. Having identified the failure modes, the steps needed to control any risks can then be prescribed. Very often this will lead to the addition of critical safety devices. Some of these devices will be passive in nature; not needed for normal operation, but needed under emergency conditions. Maintenance operations are usually production orientated, and devices that do not affect production can easily be overlooked. Unless singled out as safety critical, these devices will not command the attention warranted for their maintenance.

RCM ensures the identification of all safety-related devices and enables maintenance planning for safe and reliable operation to be based on the optimum mix of:

- scheduled maintenance;
- predictive maintenance;
- breakdown maintenance.

RCM analysis consists of:

(1) Defining the functional structure of the machine and ranking its sub-systems in the order of their effect on safety.
(2) Identifying items whose failure has a significant effect on safety and reliability.
(3) Identifying for each item the significant failure modes, their likely causes, and possible means of detection.
(4) For each failure mode, selecting the optimum maintenance procedure for minimizing the risk to safety and reliability.
(5) Setting up the maintenance plan.
(6) Implementing the plan and establishing an effective feedback loop to enable periodic adjustment and update.

6.4   Operations and maintenance

It is the responsibility of the client (or other such designated competent person as appointed by the client) to ensure that the machine system is operated and maintained so as to stay within its design limits and to ensure that the protective systems are in place and operational. This includes the maintenance of integrity through the application of appropriate operating procedures, maintenance policies, and practices. Reports and records must also be maintained in accordance with recognized QA/QC procedures for verification.

The client must ensure that personnel who operate, maintain, test, or inspect critical machine systems have received adequate training and instructions to enable them to carry out their duties safely and effectively, both for normal operation and in the event of an emergency. Emergency planning and training as required by COMAH is critical, as loss of control in the first 5 minutes of an incident can rapidly escalate to disaster. See Fig. 6.1.

In particular, process operators should be made aware of the importance of protective systems, their safety implications, and the dangers that can result from their misuse. Maintenance personnel should be aware of the importance of standards of workmanship and attention to detail in maintaining critical machine systems. Handover from a live operating plant to a maintenance activity entails a significant level of risk. A rigorous permit-to-work system, its enforcement and monitoring, are clearly necessary. To ensure that this occurs requires proper training and competent management.

The client should ensure that reports of failures of critical machine systems are sent to an appropriate designated authority who must

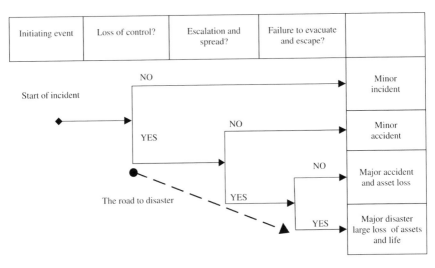

Fig. 6.1 Escalating to danger (J. Strutt, Cranfield University)

maintain a database and inform other users. The designated authority can be set up in the client's own organization and/or with the vendor as appropriate. Maintenance procedures and schedules must be routinely reviewed in the light of up-to-date information. Reference should also be made to the PUWER requirements.

6.5 Inspection

The 'responsible person', as identified in the safety file – usually the equipment package manufacturer, plant designer, or operator – will hold recommended frequencies of inspection, etc. in accordance with the Schedule of Protective Systems and Inspection Frequencies contained in the file. Pressure equipment will require a written scheme of examination by a competent person. These schedules and frequencies must be reviewed after every overhaul or examination of the machine system, and approved for adjustment as necessary (refer to the PUWER, Regulation No. 6).

Reports providing evidence that the specified periodic activity for inspecting, examining, or testing equipment or systems has been accomplished must be maintained in the equipment file, and monitored by the in-service inspection section at appropriate intervals.

6.6 System audits

Audits should be carried out to ensure that management systems that have been established to assure safe operation are adequate and are effectively maintained. The frequency of audits depends upon the process and on the machine hazards, and also on the frequency of turnover of experienced staff. In the process industries a frequency of once every 3–5 years is normally accepted as optimum for detailed system audits, although local audits by plant staff should be done more often to ensure that the actual practice follows the intended one.

Audit protocols will normally include the following checks.

- Checks on the responsibilities of key staff, ensuring that key technical staff understand their responsibilities and are trained and competent in their role.
- Checks on written procedures. Agreed practice needs to be recorded in local plant procedures so that all the staff are clear about the management systems that have been established to

ensure safe operation and the control of risks. These procedures will include any statutory requirements.

- Checks on examination and notification. The systems established to inspect and test equipment and any notification system to identify that the examination is due.
- Checks on documentation and records. The files relating to the duty, design limits, maintenance and operation requirements, and examination records.

6.7   Management of change

6.7.1   Modifications

The Machinery Directive requires that any modification must undergo the same procedure as for a new machine. Both the API and the HSE have identified the need to ensure that changes made to chemical and petrochemical plants – either in design, construction, or use – shall undergo a reasonable proving or validation process.

Modifications as applied to a chemical or petrochemical works are usually for changing, expanding, or altering feedstock, etc. and, as such, may easily result in breaching the original design parameters and concepts. Because of this, the API and the OSHA, as well as the HSE, are very concerned and aware of the possible hazards generated by such changes and have devised certain codes addressing 'the management of change'.

The aim of this section is to aid compliance with the intent of the various legislative requirements, with particular focus on the duties of the contractor or user.

6.7.2   The assessment of change

The assessment of change should cover all modifications as required by the Supply of Machinery Safety Regulations, changes of materials, changes of specification, temporary installation, by-passes, etc. which may affect the integrity of the plant or protection systems or violate in some way the mechanical or other adequacy of equipment for its specified duty. It shall also consider any changes to instrumentation, electrical, or software control systems that may affect the integrity of the process or utility operations.

6.7.3   Compliance with requirements

All changes that will involve machinery are subject to the requirements of the regulations. Any existing CE mark will thereby become invalidated, the technical file will need to be revised, and a new CE mark applied. The technical file should be updated by the user, with contact and feedback to the original manufacturer(s).

In the design and construction of plant modifications a risk assessment with a safety plan and a safety file will be required in accordance with the CMD Regulations.

6.8 Approval of modifications

6.8.1 Checks and reviews

An Environment, Health and Safety (EH&S) assessment and an essential Health and Safety Requirement (EHSR) assessment must be carried out. Whenever possible, it must be completed before any modification is carried out. If for some reason it is not possible, then as a last default, it must be completed within 72 hours of carrying out the modification. This assessment will involve deciding whether a HAZOP study or EH&S review of the modification will be necessary.

A HAZOP study or EH&S review may be carried out after the modification has been installed if the modification was required in an emergency to mitigate an unsafe situation.

Whether or not a modification requires a HAZOP study or EH&S review will depend on the opinion of the reviewers. They must decide if the modification interferes in any way with the integrity of the plant, process, or system.

6.8.2 The EH&S assessment form

The EH&S assessment form must be annotated to indicate whether a HAZOP study or an EH&S review is necessary. This form shall also record:

- engineering modifications;
- control or software modifications;
- chemical or composition changes;
- the existent and availability of safety and technical files.

This form should reflect or give attention to:

- the plant or area of change;
- the section of the plant or area, item or tag number of item;
- the date or proposed date of change;
- the details of change;
- the reason for change, and originator;
- whether EH&S assessments or HAZOP studies are required;

- problems caused by the change (or new factors that it has created) and ways taken to minimize their impact.

6.8.3 Initialling and dating

The EH&S assessment form must be initialled and dated by the following:

(1) Area facilities engineer.
(2) Process engineer.
(3) Loss prevention engineer.
(4) Control and instrumentation engineer, where applicable.

6.8.4 Recording of details

When changes are enacted, the following details shall be recorded as soon as possible.

(1) HAZOP study or EH&S review comments and recommendations.
(2) Details of the changes including process flow diagrams (PFDs), P&IDs, piping drawings, control and instrumentation details, and materials data.
(3) Safety and technical files and all documentation as required by the regulations.

6.9 Design standards

All modifications and changes should be designed, installed, and tested to recognized codes and statutory requirements where appropriate.

Chapter 7

Keep it Running

7.1 Introduction

Machinery breakdowns frequently disrupt output and affect the profitability of plant operations. It is vitally important to keep machinery running and issues of reliability and maintainability are the key to this objective. In the hydrocarbon processing industry (HPI), machinery is usually supplied to the American Petroleum Institute (API) standards. These standards have been developed over decades and represent the accumulated knowledge and experience of manufacturers and users. They are the minimum requirements for the specification of machinery covering the manufacture and procurement of designs. These requirements achieve improved energy effectiveness and reduce total life cost without sacrifice of safety or reliability; they are intended to ensure the safe shutdown of machinery before loss of containment. These standards, however, have many alternative or optional clauses that need to be carefully selected and specified by experienced specialist personnel.

Major oil companies supplement these standards with additional requirements, so as to enhance reliability of operation, based on their own experience. Alternatively, others build and accept plants to minimum standards for a low capital cost leaving the operators with a high operating cost and a struggle to improve reliability. Machinery to be installed within Europe will need to comply with the Machinery Directive and other related directives.

A more sophisticated approach is to demand a minimum life-cycle cost. Various strategies have been coined to reflect this such as:

- Design Integrity.
- Asset Integrity Management.
- TPM – Total Productive Maintenance.
- Design for Six Sigma.

They are all procedures to achieve the same objective. In different ways they focus on reliability targets at all stages in design, specification, detailed engineering, testing, and installation. The aim is to achieve the optimum life-cycle cost. This requires the co-ordinated input of research, design, development, manufacturing, operations, and maintenance. The objective is to identify and quantify losses and then find solutions for measurable gains. This then ensures that action is targeted to improve reliable operation and reduce life-cycle cost. In some plants an improvement of 1.5 per cent in reliability will provide a 5 per cent increase in profitability. Typically one day's lost production for a refinery could be worth $US 100 000 in profit. This underlines the economic need for process machinery to keep running.

7.1.1 Machine reliability

To engineer reliability, all the sub-assemblies and components that make up a machine must be categorized as:

- parts that will last the lifetime of the machine;
- parts that age;
- parts subject to mechanical failure;
- wearing parts.

Each item must then be reviewed with regard to its effect on the overall machine reliability so that the design can be revised as needed to achieve the overall objective. The maintenance actions to reinstate the machine in the event of a failure must then be established. Maintainability, which is the time to repair, has to be considered at the design stage. Designing the machine to enable a sub-assembly to be replaced could result in a shorter downtime than replacing a part. Ease of replacement rather than repair *in situ* and the logistics of spares supply become key issues. The time to failure of any component will depend on the operating conditions of the machine and how it is used. The availability of the machine to operate will depend on its reliability, which is its frequency of shutdowns and its maintainability, which is the time to repair. To ensure dependable operation all these issues must be managed.

7.1.2 Dependability management

The dependable output from any plant depends on: managing scheduled shutdowns; improved maintainability to speed turn-around; and the development of machinery reliability to avoid forced stoppages in between scheduled shutdowns. From experience, machines that have parts with limited service life that cannot work for the required duration, such as reciprocating compressors and pumps, must be provided with installed spares.

Scheduled shutdowns are usually dictated by the service life of an essential process item. Catalysts need to be renewed. Developments of new catalysts have given an extended life of 3 years instead of what was previously 1 year. Statutory inspections are being extended to match. For example, boilers with the use of on-line survey and risk assessment techniques are now beginning to achieve 6 years between shutdowns. Annual shutdowns are no longer the norm and there is a need to operate for longer periods. This places increased emphasis on plant maintenance and careful checking of critical safety and monitoring devices. Plants originally planned for a 10-year life are being extended to 20 years. API standards now require machines to be designed for a minimum life of 20 years and an expected uninterrupted operation of at least 3 years. To obtain maximum reliability and long life it is necessary to understand the characteristics of each type of machine so that the best design selection can be made to match the needs of the required process duties within the optimum operating parameters of the machine.

7.1.3 *Process hazards*

Machines are part of a process and inattention to the process requirements can be the cause of machine failure. Many processes are highly corrosive. Very often, upstream equipment over a period of time will corrode and so produce corrosion products. The process itself may produce entrained solids such as soda, chloride, borate, sulphur, polymers, catalyst powders, and so on. The damage such dirty gases can produce will depend on particle size and dust concentration. Efficient separators are needed. However, they are never totally efficient over all operating conditions and the residual dirt will have varying effects depending on the type of machine that is used.

In other cases the process produces liquids that need to be removed. During upset conditions the liquid to be separated may be abnormal for a short period. In other cases, what is normally a dry process becomes wet. In some cases process upset conditions produce corrosive elements that are not normally present. The possibility that this could occur must be verified so that the materials of construction are selected accordingly. Unless the machine has been constructed in corrosion-resistant materials even 100 p.p.m. of hydrogen sulphides and/or hydrogen chlorides will rapidly cause machine damage in the presence of liquid water.

For reliable plant operation the process design must be reviewed to identify all the events that could cause machine failure. Process safeguards to prevent the entrainment of corrosive elements, debris, and liquid carry over should be provided. In the sizing of process equipment, all operating modes together with process upsets must be taken into account. Separator hold-up facilities and liquid discharge facilities must be sized for the worst case and any process control system must be completely reliable. Even if efficient separators are installed, care needs to be exercised in their location and the design of connecting pipe runs, which should be as short as possible with no dead ends for the accumulation of dirt or liquid.

Machines on high-pressure hot gas duty may face particular hazards on shutdown due to cooling of the gas within the machine, and possible danger due to freezing and condensate or hydrate formation as the gas changes state. These pose a danger to compressor sealing systems and can cause problems due to the plugging of drilled transfer passages within the compressor, and design steps must be taken to prevent these occurrences. In a similar way, pumps on liquid service will also face hazards should the liquid form tars or crystals due to a change in temperature as a result of a process upset. Shut down of a pump on hot process duty with the subsequent cool down of its contents may also have similar problems. On the other hand, standby pumps for hot service will need to be kept at the appropriate temperature if they are to be available when required. Design facilities to maintain temperature or process shutdown procedures to avoid such problems will need to be provided.

7.1.4 Manufacturing hazards

Very occasionally manufacturing errors occur unnoticed. These are then found during operation when component failure occurs. These are typically errors in the heat treatment, resulting in coarse grain forgings or inadequate ductility of other materials. Other defects can be excessive hardness in the heat-affected zones. All these will result in stress-induced cracking or fatigue failure. Strict auditing and the application of QC/QA by qualified metallurgists are needed to avoid these problems.

Machines rely on many drilled passageways for the supply of gas or the venting of gas from seals; they likewise depend on drilled passages for the supply of lubricating oil to bearings and moving parts. In some cases internal plugs are fitted to ensure that the correct flow path is selected. Any errors and omissions can cause malfunctioning of a machine. Machines on factory test are usually at low power and these defects, if present, can escape undetected. When failure occurs in operation, to pin point the cause can be long and arduous. To check for these defects during manufacture and assembly is quick and easy with compressed air. These problems can be mostly avoided if vigilant and experienced engineers are used as third-party QA for factory inspection and testing.

7.2 Compression machinery

The type of compressor needed for any process will depend on parameters such as volumetric capacity and the required pressure. A typical compressor selection table is given in Table 7.1.

The process duty required generally determines the type of machine to be used. Where there is a choice then the hierarchy will be:

- Centrifugal – least wearing parts.
- Axial – few wearing parts but intolerant of mal-operation.
- Screws – more wearing parts.
- Reciprocating – most wearing parts.

Some typical statistics for the different types of compressor are given in Table 7.2 [see reference (1)]. As with all statistics these are average values with a wide variation from site to site and application to

Table 7.1 A typical compressor selection table

Type	Minimum inlet capacity (m^3/h)	Maximum inlet capacity (m^3/h)	Maximum discharge pressure (bar)
High-pressure centrifugal	340	50 000	850
Low-pressure centrifugal	1275	500 000	40
Axial	100 000	1 000 000	24
Screw	200	100 000	64
Reciprocating	100	10 000	800

Note. The maximum pressure shown is the available casing pressure rating. Compressor performance will be limited by the compressor inlet conditions and the compression requirements may need a number of stages of compression.

Table 7.2   Typical reliability data

Compressor type	Availability (%)	Reliability (%)	Planned maintenance (h/year)	Forced down time (h/year)	Failure rate per million hours
Centrifugal, clean service	99.7	99.8	24.8	18.8	14.26
Centrifugal, fouling service	99.0	99.5	90.6	40.6	30.85
Axial	99.1	99.8	76.4	16.4	12.54
Screw, oil free	99.0	99.5	90.0	30.0	22.83
Screw, oil flooded	97.7	98.8	199.9	99.9	76.1
Reciprocating, labyrinth piston	97.6	98.3	207.2	147.2	57.0
Reciprocating, lubricated	97.3	97.8	237.2	189.2	228.3
Reciprocating, non-lubricated	91.3	92.3	766.1	670.1	380.0

application. However, they do give an indication as to what can be achieved. Each of these types has different characteristics and is vulnerable to different hazards. To ensure average or better reliability it will be necessary to recognize the hazards and to take measures to reduce the risk from them at an early stage in design.

7.2.1   Centrifugal compressors

Operating hazards
Casings
These should last the lifetime of the machine. Elastomer components in the casing and seals are critical for reliable operation. They need to be replaced during maintenance. High-pressure service with light hydrocarbons can result in the failure of elastomer O-ring seals on decompression. This is caused by the infusion of gas at high pressure so that on decompression of a machine the residual gas pressure within the elastomer material causes the O-ring to burst. This needs to be avoided by the correct selection of the elastomer material and the review of operating procedures. Any failure of the casing from corrosion or fatigue, or failure of any joint or end cover seals on barrel compressors will result in a process gas leak to atmosphere.

Similarly, the failure of sealing of inner casings on barrel compressors can have a major effect on machine performance and efficiency.

Rotors
Rotors on clean gas service should last the lifetime of the machine. Rotors become damaged due to process reasons. Liquid slugs will cause damage to rotors. In other cases depressurizing the compressor on shutdown can result in the condensing of liquids that may cause damage on start-up if they are not drained. Debris entrainment by the process gas will erode impellers, cause mechanical damage and rotor imbalance, and affect performance. Processes that result in polymer formation in the compressor will affect performance. On shutdown, draining and solidifying of liquids on the rotor will cause imbalance with severe vibration on restarting. In the event of rotor damage, maintainability can be assured by stocking a spare rotor.

Rotor, impeller, and balance drums labyrinth seals
Dirty gas and excessive rotor vibration caused by surging or rotor imbalance will cause labyrinth seals to wear. Compression of saturated gas will cause it to become superheated as it passes through the machine. Any large pressure drop in a machine, such as across the balance drum labyrinth or across the labyrinth seal between the two sections of a rotor with a back-to-back arrangement, must be checked for any change in gas property. Damage to the rotor labyrinth seals will affect performance with a loss of compression capacity.

The pressure rise across a rotor generates an axial force and a balance drum counteracts this with the residual imbalance carried by a thrust bearing. It is important to ensure that a positive load is present for all modes of operation and that zero loads or force reversals are avoided so as to prevent any possible axial oscillation of the rotor. However, operation in surge will cause vibration and oscillation of flow, causing axial load fluctuation. Violent surging of high-power machines can result in rapid thrust-bearing failure. Any wear of the balance drum labyrinths will increase the thrust load and the thrust-bearing design margin must be able to accommodate this. Monitoring of the differential pressure across the balance drum is important in avoiding thrust-bearing failure.

Bearings
Bearing wear is a function of starting without an adequate oil film or operating with degenerated lubricating oil. Lubricating oil ages with

time due to oxidation and the action of bacterial and other contaminants. The regular testing of the lubricating oil and replacement or treatment to restore properties is necessary for long bearing life. However, bearings can be damaged due to other factors that affect their life. The use of in-line cyclone separators or other such dynamic devices with light hydrocarbon gases has been known to generate static electricity. Stray currents can cause pitting of bearings and gear teeth and earth leakage brushes will need to be installed should this occur. Similarly, all rotors must be degaussed after manufacture or overhaul.

Labyrinth gas sealing systems
This is the simplest type of shaft sealing system. Gas leakage is reduced to a minimum by the use of interlocking labyrinths between the casing and the rotor shaft so that they present a tortuous leakage path to atmosphere. These are commonly used for compressors on air and other gases where gas leakage is not a threat to EH&S.

Dry gas sealing systems
The seals are the safety-critical wearing part of the machine. They ensure containment of the process fluid. To ensure reliability, close co-operation with the seal system designer is necessary. All operating modes must be accounted for in their design. Chemical attack, bearing oil carry over, process liquid carry over, dirty gas, and Joule–Thompson effects across the seal at pressurized standstill that cause freezing, will cause failure. During operation, gas seals depend on the aerodynamic effect of a small gas leakage flow to keep the seal faces cool and apart. On shutdown, any high-pressure gas in the compressor will leak across the stationary seal faces. Depending on the gas composition, the formation of ice, condensate, or hydrate can occur. The seal faces can become glued together as a result and will remain in contact during start-up until the friction heat generated causes them to be unglued. Similarly, dirty gas from the process can flow into the seal unless provision is made to supply clean, filtered buffer gas on shutdown and start-up. Due to the need for a small leakage of gas across the seals the leakage presents a problem if the gas is toxic or flammable. This problem, and the need to prevent any uncontrolled leakage, have resulted in a number of different arrangements with the use of duplex seals. There are advantages and disadvantages with each and they need to be selected with care by all the parties concerned.

With clean seal buffer gas and continuous operation, seal life is only limited by the elastomer sealing components. Seal faces wear during starting and stopping. The rate of wear depends on how long they remain in contact. In the case of clean gas, seal face lift-off is almost instantaneous. Constant stopping and starting will affect seal life. Under ideal conditions, on clean, dry gas duty, seals should have a MTBF of over 5 years. On the other hand, seals on wet and dirty gas duty are likely to fail in half the time.

Oil-lubricated liquid film seals

These seals are becoming less common and are outdated due to the advent of dry gas seals. Here the loss of the seal oil properties due to process gas contamination and the reliability of the seal oil system are the major concerns. Consideration needs to be given to the fact that these seals depend on a controlled leakage of gas to function that needs to be flared or otherwise disposed of. There is also a small leakage of oil to atmosphere that needs collection and disposal. In some cases, this oil can be conditioned and re-used. A further consideration is that, in the event of seal failure, oil may flow into the process fluid; in many cases this is unacceptable.

Mechanical contact seals

With modern materials these seals can be made to be non-wearing. The other advantage is that the seal oil does not come into contact with the process gas and so avoids the danger of gas contamination. These seals are not suitable for high-pressure applications and are commonly used for refrigeration duties. These systems also suffer from the need to dispose of vent gas and seal oil leakage.

Instrumentation

Condensate and hydrates can also occur in stagnant instrument lines. If an instrument senses hydrates or a liquid head as well as gas pressure it will give a wrong reading. Malfunction of anti-surge control can result. Dirty gas can cause erosion of flow meter orifices; this will give rise to signal errors, which when used in conjunction with surge controls will cause their malfunction. The process gas must be checked for these hazards. Instrument lines need to be routed to avoid dead legs, and heat tracing provision to avoid condensation is often necessary.

Starting and stopping

Process considerations and providing the correct conditions for starting and stopping need close collaboration between the process designer and the machine designer. Facilities must be provided to prevent surge or compressor stonewall. During start-up the machine must preferably be unloaded for it to accelerate up to operating speed quickly and not dwell on any critical speed through which it may have to pass. The starting torque needs to be reviewed to ensure that the starting time is minimized and that stall does not occur.

For other machines that may be on hot or cold processes, start-up procedures will need to consider transient temperature conditions and the machine will need to be protected from these and any transient process effects. Machine 'settling out pressure' on shutdown and any adverse effect on seals is the other important design parameter to be considered.

Design hazards

Provided that each impeller is within its proven performance range, any slight deviation in the required performance can be corrected by trimming of an impeller blade diameter. Trimming of very narrow impellers may not be possible or may require extensive modifications. Significant deviations will require redesigned impellers. In the case of variable-speed drivers a slight change in the rated speed will avoid the need to trim impellers. Extrapolation outside proven performance has a high risk of needing a new impeller. Any changes in rotor length and weight can run into rotor instability, and care should be taken to ensure that selection is made within the designer's known technology and experience. By strict adherence to the API rotor test procedures, experienced engineers should find any instability on factory test so that the problem can be resolved before shipment. In extreme cases, however, redesign could be necessary if the selection is outside the vendor's proven experience.

Gases at high pressure have an increasing deviation from ideal gas characteristics. Gases at very high pressures behave more like liquids and large aerodynamic forces can be generated that result in the sub-synchronous excitation of rotors. Successful designs are available up to 700 bar on light hydrocarbon gases. In all cases design will depend on reliable data on gas properties.

Improving reliability

As already stated, the major effects on reliability are from the process and from inadequate auxiliary systems. Labyrinth wear will affect performance and efficiency. Apart from attention to operation and process conditions that cause labyrinth wear, the use of improved designs, such as honeycomb-type labyrinths, will be of benefit.

Vibration monitoring and balance drum pressure monitoring are usually installed to give early warning of machine failure. However, steps to ensure positive thrust-bearing loading to prevent oscillation and over-sizing for a greater safety margin will improve thrust-bearing reliability.

In the case of oil seals the installation of continuous oil treatment facilities will have a major effect on reliability. On corrosive and dirty gas operation, the use of float-type traps on seal oil drainers should be avoided and the alternative of using level-controlled drain valves should be adopted [see reference (2)].

Oil seal systems on very-high-pressure duty have a greater risk of seal pump failure. It is important to prevent damage debris from the failed pump from contaminating the standby pump. To ensure reliable use of the standby pump the use of dedicated filters for each line of pumps, instead of using a change-over valve, will prevent any cross-contamination. It will also enable damage debris from the failed pump to be cleaned from all contaminated pipework before return to service.

Dry gas seals for compressors on wet or dirty gas duty need to be designed to work in a dry, clean condition. Compressors on discontinuous duty with pressurized standstill on wet and dirty gas duty are vulnerable to seal failure. Possible measures to improve reliability are [see reference (3)]:

- to depressurize immediately on shutdown to avoid Joule–Thompson leakage effects across seal faces and to prevent dirt being carried over into the seals;
- the external supply of clean, dry, warm seal gas at all times;
- the use of heating jackets around seals and casing to avoid freezing, condensation, and hydrate formation.

For the future, encapsulated compressors with magnetic bearings without the need for seals will become common in certain applications as more experience and confidence is gained.

Finally, for machines on critical duty, especially where there is a risk of design defects, testing at full power and at the specified process duty in the factory will help to avoid problems in the field on start-up.

Characteristics of centrifugal compressors

Compressor capacity can be reduced by: suction throttling, the provision of inlet guide vanes, or speed reduction. Turn-down to about 60 per cent of design flow at constant discharge pressure, depending on the number of stages, can be provided. Further reduction in flow will cause flow instability – surge, which, depending on the power absorbed in the compression process, can cause oscillating axial forces and possible damage. Operating the compressor discharging to an empty line will result in sonic flow through the eye of the last impeller. This is overload: machines with a two-stage back-to-back arrangement could suffer thrust-bearing failure due to the reduced pressure rise over the last stage. Failure of the discharge non-return valve on compressor shutdown can result in the compressor operating as a turbine in reverse rotation if adequate recycle facilities are not provided. This can cause failure of the gas seals and bearings if they are not designed for reverse rotation. Depending on the rate of flow and the energy available from the gas, over-speed with possible damage can occur.

7.2.2 Axial compressors

Operating hazards
These machines are mostly used for the supply of large quantities of air. Atmospheric pollution will cause fouling and erosion of blades. This will result in deterioration in performance. Adequate filtration and the prevention of ice formation are critical for reliable operation.

Axial compressors have also been supplied for liquid natural gas (LNG) mixed refrigerant service and are well established in this regard. Applications on other services are unproven but the notes above for centrifugal machines would generally apply.

Machine hazards
These are in general the same as for centrifugal compressors.

Design hazards
Provided the required duty is within its proven performance range, any slight deviation in the required performance can be corrected by the adjustment of variable stator blades. Up to 40 per cent of the total number of stages may be used in any one machine. In the case of variable-speed drivers a slight change in the rated speed can also be used for correction. Extrapolation outside proven performance has a high risk of needing a new rotor design. Severe errors could result in the need to remove a row of blades. Axial compressors have long rotors possibly with many modes of vibration below the running speed. Selection of any unproven design with even a slight change is a risk. Any changes in rotor length and/or weight can run into rotor instability, which may not be easy to resolve. Any change in blade

design will change the blade's natural frequency, which must be verified and checked to ensure that excitation is avoided. In addition to aerodynamic excitation, consideration will also need to be given to the possibility of mechanical excitation from the drive train.

Characteristics of axial compressors

Due to the large number of moving blades, these machines pose a noise hazard and acoustic treatment with inlet silencers is required. Exhaust noise must also be controlled but is usually limited to acoustic, insulated ductwork with the noise mostly adsorbed by the downstream equipment.

Variable guide vanes, and/or speed reduction, can reduce compressor capacity. Turn-down of about 70 per cent of design flow, depending on the number of stages, can be provided. Further reduction in flow will cause flow instability – surge, which, depending on the power absorbed in the compression process, can cause extensive rotor damage. Operating the compressor discharging to an empty line can also cause instability due to sonic flow; this stonewall effect is critical to axial compressors and can cause severe damage. Automatic discharge pressure control is needed in addition to anti-surge control. However, this problem is avoided where the final axial stage is followed by a centrifugal impeller; the machine then has the stonewall effect of a centrifugal compressor able to handle a much larger flow. Failure of the discharge non-return valve on compressor shutdown can result in the compressor operating as a turbine in reverse rotation as with centrifugal compressors. Due to the low pressure range of axial compressor operation this is less likely to be of concern, but in some processes catalyst could be drawn back into the machine. The presence of catalyst in the machine will destroy the blading and in these situations great attention must be paid to the reliability of the non-return valve, with the possible use of a power-assisted one.

7.2.3 Screw compressors

Operating hazards
Casings

These should last the life of the machine (20 years) as specified by the API, but machine malfunction often leads to casing damage. The performance of the machine depends on maintaining the correct clearance between the rotors and the casing. The running clearance has to take into account any differential expansion between the rotors and the casing of different materials. Materials selection will be determined by the need to avoid corrosion from the process gas being compressed. Excessive operating temperatures will cause seizure as a result of loss of clearance and possible casing distortion.

In dry-screw compressors the casing has a single wall. Wet-screw compressors use the liquid to both seal between the screws and casing, and serve to cool the gases during compression. The casings are also water-cooled. In the case of water-cooled casings, any build up of deposits in the water passages will lead to hot spots and casing distortion. Keeping the water flow velocity high, as required by the API, will help to avoid deposits, but they can still occur if there are any dead pockets in the water passages. Inadequate water treatment and any hot spots could result in scale formation and possible casing distortion. Should there be any concern with regard to the reliability of cooling-water treatment, closed-loop cooling should be considered to ensure reliable operation.

Rotor seizure due to loss of clearance as a result of excessive discharge temperature and casing distortion is therefore a major hazard. This can result in the need for extensive repairs or casing and rotor replacement.

Rotors and rotor seals

For maximum compression efficiency the two screw rotors must mesh with minimum clearance between them and also between them and the compressor casing. In the case of dry-screw machines the clearances between the two rotors are maintained by drive gears. The main process gas leak paths will be across these clearances.

Although the machine is classed as a positive displacement machine, it does not dead-end and is able to accept some liquid entrainment in the process gas. Liquid slugs of course must be avoided to prevent hydraulic hammer and mechanical damage. This facility is exploited by using liquid injection to safeguard the machine from any undesirable process changes. Any polymer formation can be controlled by the injection of a suitable fluid. Hydrate and condensate can be carried through with the process gas. Temperature effects can be controlled by injection of a suitable vaporizing fluid. In the case of dirty gases, wear on the rotor seals can result, depending on particle size. In some cases, liquid injection may be necessary to ensure that the entrained dust is washed out and entrained with the discharged process gas. However, to ensure the machine is designed correctly, any process hazards must be identified during the design stage of the machine in order to provide the liquid injection and removal facilities for the package. Injection of liquid is an additional running cost and the life of the liquid could be very short depending on the process contamination involved.

As an alternative to liquid injection, wet-oil-injected (flooded) types are available. As the rotors are partially filled with oil, synchronizing gears are not required as the contact surfaces of the two rotors are lubricated. The oil is used to cool the heat of compression and is also used to control capacity by varying the quantity of oil in circulation by means of a variable inlet port. The oil then has to be separated from the gas after compression and returned for further use. As this oil is also the lubricating oil for the bearings, the oil must be compatible with the process gas with a reasonable operating life before losing its lubricating properties. Oil separation is not absolute and the resulting slight contamination needs to be acceptable to the process. Depending on the service, the oil used could be a mineral oil or a synthetic oil. Sour gas can be handled with the use of non-hydroscopic oil. Heavy hydrocarbon gas will affect the oil viscosity in a very short time. To prevent the loss in viscosity an oil-processing unit needs to be installed on line so that the oil is continuously processed to remove the contaminants. Oil life will depend on the application and replacement will range from 1 to 5 years. Both in the case of dry screws with liquid injection or the wet-oil-injected (flooded) machines the supply and disposal of the liquid involved can be a problem. Special oils may not be readily available and the disposal of contaminated liquids can pose an environmental problem.

Bearings

Hydrodynamic sleeve-type bearings and thrust bearings are used. Oil-flooded machines are usually fitted with anti-friction thrust bearings. The rotors are under constant axial load due to the differential pressure between the suction and discharge and a bending force due to the gas being compressed along one side of the rotor. Bearing life is dependent on lubrication. Oil-flooded compressors will cause the properties of the oil to degenerate due to contact with the process gas and its use as a cooling medium. This leads to a higher risk to bearing life. For dry screws the lubricating oil will degenerate due to water from the atmosphere and oxidation. Bearing life will depend on regular testing and replacement of the oil as necessary for both types of machine.

Dry sealing systems

The full range of dry gas running seals can be fitted, ranging from single seals to tandem seals depending on the application. In the case of light hydrocarbon gases, the standard design for screw machines is the application of a dual back-to-back design with nitrogen injection. There is a minute leakage of nitrogen into the process and the application is only suitable if this is acceptable. This arrangement has been well proven in service with dry-screw machines and an average MTBF of over 5 years.

Mechanical seals

These are recommended for heavy hydrocarbon gas duty where inlet pressures are low. The disadvantage is that with two rotors, four sets of seals are required so that the power losses from these seals are high. The advantage is that the seal is gas tight with no gas leakage on shutdown. Processing of sour oil leakage from these seals is required.

Oil-lubricated floating ring seals

These seals have low power losses compared with mechanical seals. Due to the high sour oil leakage the hazard is the possible dilution of the seal oil by the process gas. This can be avoided if the process gas is a light hydrocarbon. For other gases continuous stripping of the process gas contaminants in the seal oil will be necessary to keep the oil within its required properties. Security of the seal oil supply is essential at all times to prevent any gas escape.

Stopping and starting

Screw compressors do not have the flow instability problems of axial or centrifugal compressors. Due to their in-built volume ratio and the generally consequent pressure ratio, the power absorbed is a function of speed and the density of the inlet gas. Compressors operating at low inlet pressures can be started up direct on line. If the process supplies an inlet pressure greater than atmospheric pressure then the machine should be blocked in, vented down, and started on recycle. All process conditions need to be considered. Start-up with a gas composition different from design with a higher specific heat ratio will result in a possible excessive discharge temperature.

Design hazards

Design hazards are avoided by the conservative application of a standard range of casing and rotor lengths that can operate over a range of speeds. In this way any required capacity within the capacity range can be provided. By using one or two compression stages, any compression ratio up to the maximum allowable casing pressure of 64 bar discharge pressure can also be provided.

Improving reliability

Although these machines are more tolerant than other machines to process contaminants such as dirt and liquid entrainment, there are limits to what is acceptable and the process design must ensure that they are not exceeded. However, it must be noted that the smaller sizes of dry screws are less tolerant because of the corresponding smaller clearances. In all applications they must be designed in close collaboration with the process designers so that the correct type of machine is selected and any additional requirements to ensure reliable operation are identified during the conceptual design process. These requirements will depend on the inherent characteristics of the type of screw compressor selected. In other regards these machines are more sensitive to process parameters such as inlet pressure, temperature, and discharge pressure and these limitations must be established so that suitable safeguards are provided. As with all machines, maintenance of lubrication and oil properties are essential to ensure the life of bearings. The major hazard is rotor seizure and close attention to all operating conditions must be exercised to avoid over heating and the accumulation of dirt or sludge within the machine. Suction strainers should always be fitted for machine protection.

Characteristics of screw compressors [see reference (4)]

Due to the relatively large number of high-speed compression cycles being executed, oil-free screw machines are very noisy. Measures are needed for the control of radiated noise from the casing and the inlet and outlet from the machine. Noise levels from oil-flooded machines are much lower with reduced attenuation requirements.

Being a positive compression machine, it is unaffected by changes in gas properties and is the only type of compressor to handle dirty gas and/or gas with liquid entrainment. For process duties where there will be changes in gas properties, it is important to check that variations in discharge temperature due to changes in the specific heat ratio of the gas are within acceptable limits. Capacity is a function of rotor diameter and r/min. Small-diameter rotors of 102 mm are in operation up to 28 000 r/min. Larger-diameter rotors will run at lower speeds as limited by tip speed, see Table 7.3. However, the rotors are subjected to a bending force due to the

Table 7.3   Comparison of oil-free and oil-flooded screw compressors

Parameter	Oil-free screw	Oil-injected (flooded) screw
Oil contamination of gas	None	Dependent on the oil separation efficiency
Maximum discharge temperature	About 250 °C	155 °C
Maximum discharge pressure	Up to 26/64 bar	Up to 38/60 bar
Minimum inlet temperature	−160 °C	Ambient
Maximum tip speed	150 m/s	60 m/s
Capacity control	Variable speed down to about 70–50% flow	Integral capacity control system within machine down to 10% flow
Compression ratio	Limited by discharge temperature, which depends on the gas specific heat ratio Could be cooled by the injection of a vaporizing liquid acceptable to the process	Up to 20:1 depending on the gas due to the cooling effect of the oil
Dirty gas up to 300 mg/m^3	Depending on the amount and nature of the dirt, liquid injection for washing, up to 3% by volume, may be needed Liquid injection for washing will need sludge removal facilities	Not recommended. The circulation oil will need to be filtered to remove the dirt and the reliability of the bearing lubrication will be jeopardized
Wet gas up to 3% by volume	Acceptable as droplets	Not recommended. Will dilute the oil

compression effect. The rotors are therefore made as stiff as possible with the critical speed above the running speed. The maximum differential pressure allowable (at present limited to 25 bar) is a function of the rotor length/diameter (L/D) ratio. The length of the rotor also determines the volume/pressure ratio that is built into the machine. The other limiting factor is the maximum allowable discharge temperature of about 200 °C due to the possible effects of differential expansion and the need to maintain rotor clearances.

For optimum efficiency machines are designed with a volume ratio as close as possible to the required compression ratio. This means that very limited suction throttling for capacity control is possible for dry screws. The use of variable-speed drive for capacity control will enable reduction down to about 50 per cent in volume flow. As throughput is reduced the re-circulation of gas leakage across the rotor tips causes a rise in discharge temperature. The allowable gas discharge temperature therefore limits turn-down.

In the case of oil-flooded screws an almost infinite capacity control is possible by using the oil volume flow to determine the volume of gas entrained. The intimate contact of the oil with the gas is the major concern of oil-flooded machines. The application of these machines is limited to duties where the efficiency of the oil separation process is acceptable and the interface effects of the gas and the oil do not affect the reliability of the process or the machine. In the case of hydrocarbon gas mixtures it will be important to verify that operation above the dew point is maintained for all gas components during compression so as to avoid condensate and the resulting lubrication oil contamination. The maximum discharge temperature permitted for oil-flooded machines is limited to 155 °C. A summary of the two types of screw compressors is given in Table 7.3.

In summary, oil-injected screw compressors are suitable for clean gas service where a lower cost, higher compression ratio machine can be offered as compared to the use of a dry-gas screw compressor, which would result in the need for a higher cost, two-stage machine. To meet current legislation, acoustic insulation will generally be required and this should be designed to facilitate maintenance.

7.2.4 Reciprocating compressors

Operating hazards
Cylinder heads, pistons, and piston rods
The major process hazard is the carry over of liquid into the cylinder. This happens when liquid knockout drums upstream in the process are inadequate because of a process upset or because too much liquid is held up. Excessive liquid hold-up will occur if the liquid level control fails. Liquid is then entrained in the outlet gas to the compressor.

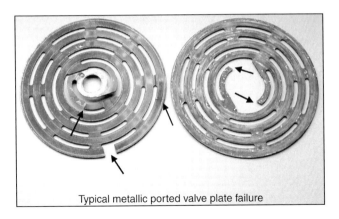

Typical metallic ported valve plate failure

Fig. 7.1   Reciprocating compressor valve

The other way in which liquid is entrained is in the compression of saturated gas. Any temperature drop of the gas on its way through pipework and pressure pulsation drums into the compressor cylinder will result in the formation of condensate. Far worse will be the accumulation of condensate into dead legs that can result in sizable slugs being entrained when they spill over.

Ingress of liquid into cylinders, depending on the quantity, will cause damage ranging from valve breakage, broken cylinder heads and studs, to cracked cylinders, broken pistons and rods, cracked distance pieces, and damage to crossheads. Ingress of liquids that are soluble with the lubricating oil can cause the breakdown of cylinder lubrication and so cause rapid wear on lubricated compressors. Liquid ingress in the case of non-lubricated compressors can disrupt the transfer films required for reliable operation.

The other major process hazard is the entrainment of dirt from the process, which can lead to valve failure, piston seizure, and the excessive wear of cylinder liners, piston rods, and rod packing.

Compressor frame (crankcase, crosshead, and running gear)
Any hydraulic forces generated in the cylinder from liquid carry over, are transmitted through to all load-carrying components of the machine. Usually the accelerometers fitted on the crankcase will cause the machine to be shut down on first impact. Components that are overloaded will have been damaged but the machine will have been protected from consequential damage due to the flailing of

broken parts and the churning of debris. The emergency shutdown of the machine will of course also limit the escape of gas from any broken gas containment components.

In the use of high-speed machines operating in the 1000 r/min range there is the added hazard of a possible crankcase explosion. Crankcase explosions can be caused by the ignition of an air–oil vapour mixture as the result of an overheated bearing or crosshead as well as by static electricity discharge. For these machines the installation of explosion relief doors with flame arresters should be considered in addition to the usual bearing temperature monitors. Oil mist concentration monitors are also available to alarm/enable shutdown before conditions reach an explosive level.

Inlet/outlet valves
These valves are wearing and ageing parts. They come in various designs:

- poppet valves, consisting of spring-loaded multiple poppets seating onto holes in the valve plate;
- slotted disc valves, slotted discs with multiple springs seating onto ported valve seats;
- concentric ring valves, concentric rings with multiple springs seating onto ported valve plates;
- strip valves.

The selection of the type of valve depends on the discharge pressures required. Generally, concentric ring valves are suitable up to any pressure, with proven experience up to 500 bar. The other types have pressure limitations. Poppet valves are considered to be reliable but they are limited to slow speed, low pressure ratio applications due to a comparatively higher valve lift needed for the same volume flow as compared with the other valves. Leaking suction valves will lose compression capacity. Leaking discharge valves will result in an increase in discharge temperature as well as a loss in capacity. Any dirt in the gas will damage valve seats and cause valve leakage. Liquid carry over with tars or crystal content will gum up and affect the operation of springs and valve plates. Lubricating oil can also accumulate due to excessive lubrication, with the same effect. Gas-entrained dirt can cause plates or rings to seat unevenly and so cause the valve springs to operate unevenly. Breakage of springs and plates, etc. can result. A MTBF of 2–3 months is not uncommon in these situations.

For improved reliability valve discs are now available in various non-metallic materials. These are used so that the valve discs can seal around any dirt as they become embedded into the material. With the use of these materials the occasional ingress of dirt can be tolerated with reported service life of 2 years or more. For optimum performance the valve lift and spring combination needs to be matched to the process gas. It is important to ensure that this is checked for each process operating condition so that adjustments to the design can be considered if necessary.

Valve covers are also subjected to the pulsating load from the compressor. In some old designs the valve cover/valve cage gaskets suffer plastic deformation and result in the fatigue failure of the holding down studs. The blow-out of valve covers is a safety hazard and results in the loss of process gas containment.

Piston rings, wear rings, and cylinder liners
Entry of dirt into cylinders will cause the accelerated wear of piston rings and wear rings, and possible scuffing of cylinder liners. Entry of process liquids that are soluble with the lubricating oil used will result in the dilution of the lubricating oil properties leading to accelerated wear. The wear rate of thermoplastic piston rings and wear rings is affected by temperature. Any loss of cylinder capacity at one stage will result in an increased compression ratio at other stages and so increased discharge temperature. Leaking valves will also cause an increase in discharge temperature, as can extended running on part-load; this is due to recycling of hot gas. Any of these events will result in an increase in operating temperature and so cause rapid wear. Rapid wear, if undetected, will cause the piston to come into contact with the cylinder liner with a possible seizure and major damage to the machine. Consequently, cylinder discharge temperatures need to be monitored and alarmed together with rod drop alarms for each piston rod.

Rod packing
The wear rate will depend on the pressure, the process gas, the surface finish of the rod, piston speed, and packing material. The need for lubrication and cooling of the rod packing will generally be in accordance with the rules laid down by API standards. Due to the relatively small dimensions of the cooling passages the use of utility cooling water will lead to blockages and overheating with the resulting increased wear. While the use of filters will help to prevent blockage, the use of a closed-loop cooling system with an

enhanced heat transfer fluid should be considered for improved reliability.

Packing failure will lead to the loss of containment and a loss of efficiency. However, the use of vented and purged packing should ensure that the loss of containment is controlled.

Stopping and starting

Due to their geometry, crankshafts are torsionally flexible. During start-up, a number of torsional critical speeds will need to be passed on the way up to the running speed. Reciprocating compressors must be unloaded as far as possible to avoid the build up of fatigue cycles. This will be especially true in the case of synchronous motor drives with the danger of their oscillating torque being coincident with a system torsional critical speed during run up. In these situations it is important to know that in the case of multi-stage high-pressure machines, venting down to the stage suction pressure will not give the lowest start-up torque. Gas forces on each side of any double-acting piston are not balanced due to the presence of the piston rod. For minimum torque it will be necessary to vent down all stages to the first-stage suction pressure.

Stud bolts and tie rods

These items secure all the critical components of the machine that carry the forces of compression. They secure the crosshead guides to the crankcase, the distance pieces to the crosshead guides, and the cylinder assembly to the distance piece. In any major disassembly, strict adherence to assembly instructions and the correct torque loading are needed to ensure the uniform distribution of loads. In some designs tie rods may be involved that need to be regularly torque checked.

All these items are subjected to vibration and cyclic stress; incorrect assembly can lead to mal-distribution of loads resulting in fatigue failure, otherwise they should last the life of the machine.

Crossheads

These carry the side loads due to the articulating motion of the connecting rods. They are usually fitted with replaceable shoes. The use of conservative loading and adequate maintenance of the lubricating oil should provide long life. However, damage will occur should there be a major overload as caused by the ingress of a liquid slug in a cylinder.

Design hazards

Torsional failure

Reciprocating compressors are usually designed using standard components that are well proven. The design hazards unique to each application will be torsional fatigue due to the pulsating nature of the compression process. The crankshaft driver train has to be designed to reduce torsional oscillations within acceptable limits by the correct sizing of a flywheel. This design procedure is well established and with due care the risk from this hazard can be minimized. With the use of variable-speed drivers the permissible speed range is restricted because of the need to avoid critical speeds, and safeguards are necessary to prevent any adjustment of speed stops by operators.

Vibration-induced failure

By the very nature of the machine, all components on or connected to the machine will be subjected to vibration and all gas-containing parts will be subject to gas pulsating flow. Usually analogue studies are carried out on the process piping system connected to the machine and the machine stages are supplied with pulsation dampening vessels to limit the resulting gas pulsations. These studies do not, however, prevent the forced vibration of the compressor pipework and attachments. Very often, local pipework can suffer fatigue: instruments, pressure relief valves, and other valves will become unreliable. These need to be carefully reviewed and checked for the likely induced vibration. This is especially true in the case of large machines where the residual gas pulsations result in significant forces due to the larger-diameter pipework and component weights involved. It should be noted that the API rules for pulsation control were primarily formulated to ensure compression efficiency. It is equally important to verify the gas forces generated on large machines and the effects that they have on all items of equipment. Reciprocating machines are usually installed with a spare to ensure reliable operation. Very often it is overlooked that a routine change over of machines is needed for maintenance purposes. In order to do this without affecting the process, it will be necessary to transfer load from one machine to another. If this results in two machines operated at 50 per cent load for a period of time it may be that the pressure pulsations for this operating mode are unacceptable. This shows that all operating and transient modes of operation must be considered.

Piston rod failure

The available cylinder sizes and more importantly the available rod loads are the design limitation on each compression stage. The piston rod is one of the most critically loaded parts of a compressor, it is subject to fatigue failure, and its life is dependent on its adequate preloading. In the past, inadequate designs and unreliable assembly techniques have resulted in failures so that many users required increased safety margins to avoid failure. Current designs adopt measures to avoid stress raisers and use methods to control accurately the pre-tension of the piston rod assembly to the crosshead and piston. In other cases the use of a flanged attachment of piston rod to crosshead has been adopted so that the attachment load is shared between the rod and flange due to a sort of compound bar effect. This then results in reducing the stress concentration in the rod attachment in order to avoid fatigue failure [see reference (5)].

Another design hazard is the possible overload of the piston rod due to alternative operating conditions. One of these conditions is overload. As reciprocating compressors are positive displacement machines, each stage has to be protected by discharge pressure relief valves (PRVs). Rod loads must be checked for operation at the PRV set pressure (usually 110 per cent). Another condition is part-load. The load on a piston rod is made up of two components. Firstly there is the inertia load from the piston at top- and bottom-dead-centre as it comes to rest and changes direction. Secondly there is the gas compression load due to the differential pressure across a double-acting piston. This gas load acts against the inertia load and reduces the actual rod loading. Unloading one end to reduce capacity will change the rod loading. However, changing the duty to lower the compression ratio or to discharge at a lower pressure will also change rod loading. In some cases users may think that this results in a less arduous duty whereas, in fact, rod loading has increased. This is especially true for high-speed machines where the inertia effects may be more dominant. It should also be noted that the lubrication of the connecting rod little end depends on load reversal for the influx of lubricating oil. Compressors are designed to enable this and again operating outside of specified operating conditions can cause failure of connecting rod bearings. Designers cannot ensure machines are designed correctly unless all possible operating modes are identified at the design stage. Any change in operating conditions without a proper design review runs the risk of machine failure.

The design of piston rods is further complicated by the need to seal the rod from the escape of process gas. The action of reciprocating motion of the rod relative to the rod packing causes wear between the two surfaces. The rod material, in order to resist fatigue, must be reasonably ductile, but the rod surface needs to be hard to resist wear. The material if in contact with corrosive gas must also comply with NACE requirements.

Maintainability of the rod packing is another concern. One option is to design the rod to allow the installation of pre-assembled packing. This requires that the rod is disconnected to enable this to happen. The other option is to design the packing for disassembly and replacement of the worn parts with the rod in place. This can be tedious and prone to error, especially in the case of restricted access. These options need to be carefully assessed for each application and size of machine.

Tribology hazards

It has been well established that valves, piston rings, rider rings, and piston rod packing have a large impact on compressor reliability. Reliability is dependent on the correct selection of materials so that the boundary conditions between the rubbing materials are lubricated correctly in the presence of the process gas environment. With the availability of PTFE over the past few decades and the more recent arrival of high-duty polymers such as PEEK, it has been possible to operate reciprocating compressors without the need for cylinder and rod lubrication. For enhanced reliability the use of non-lubricated compressors is preferred where there is proven experience for the proposed application. The use of an oil system is troublesome, unreliable, and costly; it affects the process and in many cases the resulting contamination is not acceptable.

Non-lube design

Incorrect material selection will result in poor reliability, with failures after too few running hours. PTFE and PEEK are plastic polymers that have low coefficients of friction but they have to be combined with fillers to improve their mechanical properties. Typical fillers are:

- carbon graphite;
- fibreglass;
- metal powders;
- ceramic powders.

So-called polymer alloys, which are a mixture of polymers and fillers, have been developed so as to provide a range of physical properties that provide good wearing properties to match the various process operating conditions that may be encountered. These have tensile strengths ranging from 9 to 70 MPa with Shore 'D' hardness ranging from 60 to 90. The fact that such a range of materials needed to be developed demonstrates the complexity of selecting the correct material for any given application.

The success of using polymer alloys in non-lubricated applications depends on the chemistry of depositing a film of polymer on to the surface with which it is in sliding contact. An adequate deposit depends on the surface finish and the presence of an oxidized film and any chemical reactions from the gas being compressed.

The parameters that affect material selection are:

- the degree of moisture present in the gas and its dew point;
- the presence, or non-presence of oxygen;
- the type of process gas and whether or not it decomposes due to the heat of compression and/or the heat generated as a result of friction between the rubbing surfaces;
- the differential pressure to be contained;
- the operating temperature at the material surfaces.

The operating limits for the use of polymer alloys in non-lube service are:

- 350 bar discharge pressure;
- a mean piston speed of about 4.0 m/s;
- a mean temperature in the range of 120–175 °C depending on the polymer alloy used;
- avoidance of wet gas as it affects reliability and ring life.

Gases with entrained process fluids cannot be tolerated due to the possible development and accumulation of sludge. The sludge could be abrasive or have a chemical effect on the polymer film on the counter surface, which will lead to a rapid increase in wear. Operation outside of these limits will require lubrication [see reference (6)].

Lubricated design

The design requirements for compressors with metallic piston rings and packing are well established but they are usually avoided due to the need to use significant quantities of lubricating oil. The use of mineral oil poses a fire hazard, especially in the compression of an oxidizing gas such as air. If discharge temperatures become high any oil sludge that has built up will ignite. The oil fouls up the process and in other cases contaminates the product. In some cases synthetic oils are used to avoid these problems.

The use of polymer alloys with lubrication allows the applications of these compressors to be extended. The oil quantity required for effective lubrication is far less than for metallic materials, hence the term 'mini-lube'. The operating limits for mini-lube operation are increased to:

- 1000 bar discharge pressure;
- about 200 °C discharge temperature depending on the polymer alloy used;
- allow operation with wet gas provided the lubricant grade and feed rate are adequate to overcome the dilution or washing effect of the entrained liquids.

The quantity of oil injected is critical: too little can cause sludge build up and too much could result in oil accumulation in the cylinder, which will affect valve operation and compressor performance. The quantity of oil needed is a function of the swept surface area, the operating pressure, and the solubility of the oil with the process gas. This means that the lubrication rate should be different for each compression stage. Hydrocarbon gases are more soluble with mineral oils and may benefit from the selection of synthetic oils. Higher-pressure stages need higher viscosity oil and usually a balance has to be made between the needs of the lower-pressure stages and the high-pressure stages. Oil needs to be supplied so as to ensure that the wear rings run on a wetted surface and the minimum number of feed points will depend on the size of cylinder and, in the case of rod packing, the operating pressure. Typical requirements are given in Table 7.4.

Improving reliability

The machine selection should be made with a low mean piston velocity compatible with cost. The availability of the machine will depend on the avoidance of liquid and dirt ingress. Where the compression of saturated gas is involved condensation within the compressor installation must be avoided. Depending on the location of the process knockout drum there can be a temperature drop in the gas arriving at the compressor package so that condensate is formed. In these situations additional provisions should be made such as:

Table 7.4 Typical numbers of feed points

Cylinder bore diameter (mm)	Minimum number of feed points	
	Cylinder	Packing
Up to 254	2	1
254 to 610	3	1
610 to 864	4	1
864 and over	5	1

For cylinders operating at 100 bar and above, a minimum of 2 packing feed points is needed.
Packing feed points should always be located at the top.
Cylinders should be lubricated with one upper and one lower feed point. Any additional cylinder feed points, as indicated above, should be located towards the top to ensure complete lubrication of the swept areas.

- ensure that cooling water to the cylinders is above the inlet temperature of the gas by at least 5 °C;
- provide heat trace inlet pipework and pulsation bottles, liquid knockout facilities, and blow-down facilities and avoid dead legs;
- ensure that interstage cooling avoids the production of condensate; otherwise ensure the adequate installation of interstage knockout drums and demisters.

It should be noted that machines are usually designed with the lowest possible interstage inlet temperature for maximum efficiency. When this results in condensate being formed, it may be a better option, where possible, to adopt a higher interstage inlet temperature that avoids condensate. This then avoids the need for separators and reduces the interstage pressure drop and so offsets any loss of efficiency. However, condensate may still occur as a result of a change in gas composition or a process upset condition and this must be checked. If this should be the case then separators must be provided as a safeguard. The avoidance of corrosive condensate is a further bonus. Note that in the case of gas mixtures the condensing temperatures for the various components will depend on their partial pressures and this will need to be analysed by process simulation.

In the case of lubricated machines there is a deliberate injection of oil but entrainment of liquid with the process gas must be avoided. This paradox shows that the injection of lubricating oil must be carefully metered and the reliability of the oil metering system is of prime importance. It should be noted that feed rates are measured in drops per minute. To prevent back flow a system of non-return valves

Table 7.5 Checklist to ensure reliable operation

Possible defect	Action needed to check
Lubricator line check valves leakage and possible faulty lubricator oil plunger	Any defects will result in the presence of process gas. Check each lubricator sight glass for this.
Lubricator plungers	These are subject to wear and need regular inspection and maintenance.
Sight feeds	Check that the oil level is below the drop feed pipe to ensure that drop feed rates can be properly monitored. Adjust the level by unscrewing the sealed plug at the top of the sight glass and allowing the oil to be pumped out.
Feed rates	Check daily the feed rates (in drops per minutes) to each point. Correlate to the total oil consumed from the reservoir as measured by the quantity of top-up required and the time interval in minutes. This needs to be based on the equivalent drops per litre for the type of oil used.

has to be fitted to each feed line. Due to machine vibration, drift and component failure is common. Redundancy should be provided and routine maintenance should be carried out in accordance with Table 7.5.

As explained above, reciprocating compressors suffer a loss of capacity as a result of the deterioration of valves and piston rings. This will result in other stages working at higher ratios with a general increase in discharge temperature. This will lead to increased wear. To improve reliability and to avoid changes to the interstage pressures and temperatures, critical machines can be specified with say 10 per cent excess capacity. By using clearance pockets or by-pass, the machine capacity can be reduced to match the design requirement. In the same way, any loss of capacity can also be accommodated so that discharge temperatures are maintained.

Reciprocating compressor characteristics

Reciprocating compressors are designed to operate in the conditions specified by the user. Operating under any other conditions can cause machine failure. As the design centres on limiting rod loads and operating temperatures, operators need to be trained to avoid operating conditions that have adverse effects on the machine.

Even operating with cylinders unloaded for any length of time can cause damage. This can cause overheating of cylinders and accumulation of oil with oil-lubricated machines, among other problems that could occur. Commonly, insufficient attention to start-up, load transfer, and other transient operating procedures can give rise to problems of premature failure.

7.3 Centrifugal pumps

Operating hazards
Process fluids
Pumps are used to move a variety of process fluids. Some are benign and failure of containment is just a housekeeping problem. Others pose a hazard to EH&S and total containment is mandatory. The fluids can have properties such as being corrosive, or being very hot, or very cold. Pressurized hot water, for example, will become superheated steam when released to atmosphere.

The process hazards to pump operation can be entrained debris, or a change of state of the fluid associated with a change in temperature or pressure. Fluids that can become solid while pumping form crystals or polymers that will cause serious damage. In some processes this can occur during upset conditions or as a result of unreliable process controls. Cold liquid gas can easily change to vapour at ambient temperature. Liquids that can become vapour while pumping are another danger that can result in damage. Variations of the process temperature will affect the pumped fluid viscosity and may affect performance. All these different problems need to be identified during design to ensure a reliable pumping installation.

Fig. 7.2 Centrifugal pump impeller inlet vane showing cavitation damage

Casings

The material standards for all expected process fluids are well established. They are covered in API specifications and there should be no problem in achieving long and reliable life for the casing. This of course is not the case for slurry pumps and special provisions will be needed that can cost up to three times more to avoid wear.

Rotors and wear rings

Impellers, wear rings, and throttle bushes may suffer erosion and corrosion effects due to upstream debris entrained with the fluid. Process fluids that may form polymers or develop crystals will likewise cause damage. Wear rings provide the sealing between the impellers and the casing. Any increase in clearance caused by erosion and impeller rubs will affect the pump performance due to increased leakage recirculation. Damage to impellers or the build up of polymers will affect rotor balance resulting in increased vibration and damage to mechanical seals.

Depending on the fluid density, damage can also be caused by cavitation when operating with insufficient net positive suction head (NPSH). This is caused by operation at a pressure below the vapour pressure inside the pump so that gas bubbles are released at the impeller eye. These then collapse as the fluid increases in pressure through the impeller. Any increase in recirculation from the first-stage impeller discharge back into the impeller eye could affect the required NPSH. Recirculation of fluids with high specific heats, perhaps three times that of water, could affect the vapour pressure at the impeller eye.

Bearings

Pumps are fitted with anti-friction bearings or hydrodynamic bearings depending on the speed and kilowatts transmitted. The journal bearings that carry the rotor will be affected by excess vibration. Thrust bearings carry the hydraulic unbalanced axial forces caused by the pressure rise across the machine. The reliability of bearings depends on adequate lubrication but excess vibration due to imbalance or flow-induced hydrodynamic forces may cause damage. Increased wear ring clearances that affect axial balance, depending on the balancing arrangements, can in some instances overload the thrust bearing and reduce its operating life.

Shaft seals

The most common cause of pump failure is the loss of containment due to seal failure. It has been suggested that up to 64 per cent of pump failures are related to seals. The possible problems of seal reliability can be caused by:

- incorrect operation;
- incorrect installation and maintenance;
- bad piping design/pump application;
- incorrect seal selection.

Seal systems fail when they are not compatible with the process fluid. Any lack of proper lubrication across the seal faces will cause failure. To avoid failure they must be designed to meet all possible variations that may occur in a process. Seal system design is therefore the key to seal system reliability and it is vital that all parties are involved in the design. That is:

- the seal manufacturer's design engineer;
- the pump design engineer;
- the process design team.

In the case of low-pressure water utility pumps, ordinary packed gland seals can be used. For most other applications the use of mechanical seals are the norm and their characteristics need to be understood.

Seal running clearances are in the order of microns. Large particles in the sealing fluid will accumulate around the seal and restrict its movement and cause hang up. Fine particles will pass through the seal and damage the sealing faces. To prevent this a clean flushing fluid needs to be injected into the seal cavity as a barrier between the seal and the dirty process fluid. Seals produce heat due to friction. Cooling facilities are needed to carry away the heat.

To prevent seizure, seal faces need boundary lubrication, therefore they leak. In general the leakage quantity is so small that in most cases it will vaporize due to the heat of friction without leakage being apparent. For fluids hazardous to EH&S, even this small amount of vapour may not be acceptable.

The fluid that leaks across the seal faces will drop in pressure and increase in temperature. If this causes a change of state such as the formation of crystals or carbonates, this will accumulate at the exit region of the seal and eventually cause the seal to stick. In order to prevent this an auxiliary quenching fluid has to be used.

The seal system design requires a match between the process fluid properties and the need to meet the seal operating conditions as given above. There are three types of seal and three alternative arrangements with a diverse range of piping plans depending on the process operating conditions and the need to alarm and contain the process fluid in the case of seal failure. Probably the most complex of these sealing systems are the dual-seal arrangements. These are required for fluids hazardous to EH&S. A dual-pressurized-seal arrangement is needed when there is a:

- low vapour pressure margin;
- process fluid such as carbon dioxide with unacceptable lubrication properties;
- corrosive fluid;
- dirty, abrasive or polymerizing fluid.

The API piping plans deal with the sealing support systems and the supply of external fluids. They show the arrangements needed such as valves and instrumentation for control and monitoring. It should be noted that attention to operability and maintainability are just as important as selecting a suitable sealing system.

A procedure for selecting a sealing system is given in API 682 (ISO 21049), second edition 2002 *Pumps–Shaft Sealing Systems for Centrifugal and Rotary Pumps.* See Annex A, Recommended seal selection procedure.

Improving reliability
Verifying design assumptions
In addition to the various tests that are required as listed in the API standards the following additional requirements should be considered as appropriate.

Dual-pressured seals are supplied with a barrier fluid that leaks through the inner seal to the process and through the outer seal to atmosphere. Barrier seal pressure is required to be slightly higher in pressure than the pump seal cavity in order to minimize its leakage. The seal cavity pressure is usually estimated to be 20 per cent of the pump head or some other empirical formula. As the barrier seal pressure is important for the reliable operation of the seal, it is a small matter to verify the pump seal cavity pressure during pump test so as to avoid any possible error.

Pumps have unbalanced hydraulic forces that need to be carried by a thrust bearing. There is a danger that the axial thrust may become zero at some point in the operating envelope. This can lead to undesirable axial oscillations. This should be checked at the design stage so that the required thrust-bearing capacity is also confirmed. Monitoring the shaft position during pump test will verify that axial force is always in the same direction.

The use of seal-less pumps
Single-stage seal-less pumps are available that are suitable for many petrochemical applications and avoiding all possible leakage of fluid. The rotor and bearings are contained in a completely sealed enclosure, apart from the flanged connections, and the rotor is driven via a magnetic coupling. In other canned designs the motor is submerged in the process fluid and directly drives the pump. Another alternative design is where the motor rotor is encased with the process fluid with the motor stator outside. In all these designs, while the need for a sealing system is avoided, reliability depends on safeguarding the lubrication of the bearings by the process fluid. They are therefore still vulnerable to the normal hazards from the process. However, pumps have become more reliable due to the development of bearing materials with design features that will accept the presence of small quantities of particles, up to $100\,\mu m$ in size. In other cases the provision of a filtered product stream into the bearing chamber will be needed. Lack of lubrication can occur in the case of pumping cold liquids that will warm and flash to vapour on shutdown and special safeguards are needed to ensure that the pump is fully primed for start-up.

Stopping and starting
Pumps on hot service may need special shutdown and start-up procedures depending on the fluid. This may be especially applicable for multi-stage pumps with long rotors. It may be necessary to maintain operating temperature to prevent thermal distortion. Other procedures may be required to keep differential temperatures within limits, depending on the pump design.

Pumps on process fluids may need provisions for isolation and the discharge of the inventory with flushing facilities. This will depend on the nature of the fluid. It is important to verify any change of state of the fluid as a result of temperature and pressure changes. For example if the fluid solidifies any attempt to start the pump could wreck the pump and driver. Drivers need to be adequately sized to allow start-up under all possible conditions. All operating requirements – such as commissioning with an alternative liquid, first-fill conditions – should be checked to ensure that the pump driver has sufficient power for all these conditions.

Part-load operation

To cope with a range of part-load operations the use of a multiple pump installation is usually adopted. In most cases spare pumps are also installed. Start-up and by-pass facilities need to be sized to handle all transient flow conditions of excess capacity. After extended operation, the performance of the pumps may have been reduced due to increased internal leakage as a result of impeller damage and worn casing wear rings. With a loss in capacity and a desire to defer maintenance, operators will resort to running the spare pump to make up the needed capacity. It is possible that this can result in pumps being operated in an unstable region. This needs to be checked and the capacity control system adjusted for the changed pump characteristics as necessary so that existing damage is not compounded.

Design hazards

Design limits

Pumps selected within well-proven parameters should not give rise to problems. Multi-stage pumps designed for high pressure and operating at speeds above 4000 r/min need special design features to avoid rotor stability problems. They must be supplied with devices such as swirl brakes and gap control to prevent the development of the hydraulic forces developed by secondary flow that can cause rotor instability. To ensure reliable operation the supply of these pumps needs to be restricted to those who have the requisite experience. These pumps also have large axial forces and two-stage designs are adopted so that the axial forces can be balanced out. The need for long-rotor multi-stage pumps is cost driven as the adoption of a two-casing selection will lead to a higher cost. In some applications it may well be more prudent to use a two-casing solution for greater reliability.

Cryogenic pumps

Pumps on cold liquids such as liquefied gases will need to be cooled down to pumping temperature for start-up. Cool down is done by opening up the isolation valves and blowing down the casing until the operator thinks that the correct temperature has been reached. In the case of cryogenic pumps, frost and ice will soon cover the casing. The operator is expected in some cases to recognize by the change in colour of the blow down when the right temperature is reached. To rely on the operator will lead to errors and damage to pumps. The provision of temperature monitors on the casing will help to prevent operator error but this alone will not indicate the presence of liquid because saturated vapour will have the same temperature and some time interval will need to be allowed to be sure that liquid is present. The use of liquid sensors will overcome this problem. It will also be important to provide vent-off facilities arranged to ensure that no gas pockets are trapped within the pump. The same problems will apply for in-tank pumps during commissioning or re-installation after maintenance.

System resistance curves

Duty points are specified for pumps but system curves should always be developed to ensure that part-load duties are properly identified. This will especially be the case for multiple-pump installations. All operating modes should always be established and the system matched against the pump characteristic to ensure operation within the allowable pump envelope. It is important to ensure that the pump head rise from the duty point is steep enough to ensure a definitive intersection with the system curve in order to avoid any instability in operation. This will be especially true for constant-head systems that are unaffected by flow.

System curves that are dominated by the head generated from flow resistance will be exponential in shape. With the need for a multiple-pump installation the rated duty point with all pumps in operation is likely to be at the pump best-efficiency point; at part-load with less pumps in operation the pump operating point will most probably be to the right of the best-efficiency point towards overload. This means that the pump required NPSH will be increased dramatically and therefore this needs to be verified and catered for accordingly.

Very often safety margins are built into system resistance curves in order to allow for fouling and corrosion. Sometimes this can lead to pumps being over-sized for the as-new-condition service. Pump operation must always be checked for all conditions of service to ensure compatibility.

Pump characteristics

Centrifugal pumps have an allowable operating range around the best-efficiency point based on the design flow and head. The developed head at the best-efficiency point increases as the flow is reduced until a point of instability is reached. This is the minimum flow point of stability. Further reduction in flow will reach a point of thermal instability. On the other hand, the developed head will decrease as the flow is increased from the best-efficiency point. This is due to the fact that turbulence losses increase as a square of the flow. Depending on the impeller design characteristics the increase in flow, well beyond the best-efficiency point, will eventually reach a point where all the energy is dissipated into heat as a result of turbulence. Pump operation should be limited well short of such a point. The power required increases from minimum flow to overload and usually drivers are sized from 110 per cent up to 125 per cent of the required power at the duty point, depending on the size of the pump. The required NSPH also increases with flow from the duty point due to the pressure losses increasing as a result of turbulence.

While pumps are used to convert kinetic energy to pressure it is as well to recognize that pumps will equally convert pressure energy to power output and can be used as hydraulic turbines. Care must be taken to prevent reverse flow in pumps that could result in pump over-speed and possible damage due to the resulting reverse rotation.

7.4   Drivers

7.4.1   The machinery train

The process machine train consists of the process machine, the coupling, and the driver. A train consists of elements in series; the failure of any one item will cause the unit to fail.

One critical item is the coupling that connects the driver to the process machine. The relative position of the driver and the driven machine will be affected by thermal expansion. This can be the result of the temperature differences between night and day in the tropical sun or the change in temperature of the machine during operation. Depending on the size of the machines and the degree of temperature change there can be significant changes in the machinery alignment. Usually attempts are made to ensure alignment in the normal operating position. In other cases there may need to be a compromise. The technology used for machine alignment can range from the use of mechanical dial indicators to the use of laser beams. In order to ensure the correct selection of the coupling and the co-ordination of the rotor dynamics of the machinery train unit, responsibility should be assigned to one vendor.

Early designs of couplings, many of which are still in operation, were of the gear type with the angular misalignment limited by choosing the correct distance between shaft ends. These designs depended on the spacer shaft being allowed to articulate in order to accommodate any misalignment. Lubrication is critical to their reliable operation. If any sludge or dirt is allowed to accumulate in the teeth, lock-up can occur. An increase in bearing loads will occur and there is a risk of a fire if the oil vapours that are produced are ignited.

Modern couplings avoid these problems by the use of flexible elements with different problems. Excessive misalignment will give rise to fatigue failure. High-strength flexible elements are vulnerable to chloride corrosion attack from sea-coast atmospheres. The relatively larger diameter of the flexible elements generates heat due to windage effects. These risks need to be recognized so that action is taken to safeguard against them at the design stage.

Gearboxes are often needed when low-speed drivers are used with high-speed compressors. These may give rise to tooth passing frequencies that cause excitation of the blading when steam turbines or axial compressors are involved. This possibility should be verified at the design stage.

Some of the many types of driver that can be involved in a process machinery train are discussed in the following sections.

7.4.2 Special-purpose steam turbines

Process plant may need several levels of steam pressure and the steam system will be designed to supply many users. Special-purpose steam turbines are designed for many different roles in such a system: operating as back-pressure turbines, as pass-out turbines, as condensing turbines, or any combination of these. They are usually of modular construction made up of matched standard elements to meet the required duty. They are multi-stage machines and those that work off the main steam header from a boiler installation are exposed to hazards from the steam supply that will affect their reliability. Other steam turbines that are supplied with steam indirectly from the boiler will suffer the same corrosion effects but will be less likely to suffer from deposits, as these will have mostly precipitated out elsewhere.

Process hazards from steam

The production of steam is a process and in common with any other process more attention is needed in the design of the process to improve machine reliability.

Fig. 7.3   **Deposits on steam turbine blades**

Steam is produced from water, which is not pure. On start-up the boiler feed-water is from a demineralized water supply contained in storage tanks. The storage tanks are filled at a constant rate from a water treatment plant that needs regeneration at periodic intervals. Although this treatment will almost purify the water some minute residuals remain. The treated water, however, can also be subsequently contaminated, perhaps by rainwater leakage into the storage tanks or by the supply of poor-quality water due to operator error. Water also absorbs oxygen and carbon dioxide gas. The solubility of these gases is a function of their partial pressure and temperature. To remove these gases it is usual to use a pressurized degasification process (a de-aerator) in the feed-water supply to the boiler. Chemical dosing is then used to deal with any remaining oxygen present and if necessary to correct the pH value.

During normal operation the feed-water to the boiler will be made up of condensate return and demineralized make-up water. In order to achieve maximum cycle efficiency, steam is condensed under vacuum. Invariably air will leak through any joints under vacuum and although air is ejected from the condenser some oxygen will remain in the condensate. The operation of the degasification process will be affected by the proportion of condensate and demineralized water in the feed-water to the boiler. The returning condensate will be less than the steam generated as a result of steam losses in the system.

The feed-water in the boiler is converted into steam but any remaining solids or salts will gradually accumulate as the water is converted into steam. To control this, some of the contaminated water has to be discharged and replaced by the purer feed-water. The steam generated in the steam drum is saturated and is passed to the superheater where it is raised in temperature so as to vaporize any water droplets carried over and to produce dry steam. The boiler design will normally incorporate separators in the steam drum to minimize the water carry over. Any solids or salts carried over with the water droplets will remain and are entrained with the steam. The steam will also contain any oxygen that has not been removed. It is the adverse effects of these steam impurities that affect steam turbine reliability and availability.

The process of ensuring steam purity as outlined above is a complex one and its efficiency can be affected by off-design conditions such as:

- start-up;
- part-load;
- cyclic load;
- abnormal steam loss.

All the above conditions will affect boiler blow down, chemical dosing rate, control of the degassing system, and the efficiency of water separation in the stream drum.

Part-load operation of steam condensers has a serious effect on their performance, as will variations in cooling-water temperature. This results in lowering the condensing temperature of the steam and its vapour pressure. Any increase in vacuum will increase air leakage, which will cause the partial pressure of the air in the condenser to increase if the condenser vacuum system does not respond to the changed conditions. Unless part-load conditions are considered, condensers designed for the rated load will have under-sized vacuum systems for part-load operation that can result in highly corrosive condensate being returned to the boiler. This presents a corrosion hazard to the turbine section under vacuum, the condenser, condensate pumps, and associated pipework. In order to avoid these problems the vacuum control system of the condenser must be adequately sized and designed for part-load operation.

Pressure degasifiers depend on temperature to perform. The temperature control of the process must respond to any variations in the returning condensate to ensure the required oxygen removal.

The quantity of chemical dosing must also be adjusted. In some cases the use of an alternative vacuum degasification process could be more reliable as the recommended oxygen levels of the condensate can be assured without the need for chemical dosing.

In consequence steam purity will be affected by these conditions and the whole process of removing impurities should be reviewed for all required operating conditions to ensure turbine reliability.

Operating hazards
The effect of poor steam purity
Steam turbine long-term reliability and availability is seriously affected by impurities in the steam. Oxygen in the steam will corrode the steam circuit and the entrained products of corrosion will impact and erode nozzles, blades, and guide vanes. The production of iron oxides will also form deposits on blades. Silicic acid and silicates form deposits that are hard to remove. The presence of CO_2 will form a corrosive steam impurity. These and other contaminants such as sodium, hydroxides, chlorides, sulphates, copper, and lead will all shorten turbine life. Failure can occur as a result of corrosion and stress corrosion cracking, and turbine blades can be fractured. Erosion and corrosion pitting and deposits can cause rotor imbalance with increased vibration and the fatigue of bearings. Continuous operation at the same load point can result in the steam inlet governor valve being locked in one position because of the build up of deposits. Regular exercising of the valve can prevent this. In less severe cases the turbine may well run reliably until shut down for routine inspection. On inspection internal damage will then be found. Rotors may need to be extracted for the removal of deposits. Any cracked blades found will need to be repaired or replaced. Other static components may also be found to be in need of repair. A prolonged shutdown is then necessary with a severe effect on availability.

Impurities in the feed-water have to be limited in order to avoid boiler failure but the steam quality acceptable to boilers and steam pressure systems may not be adequate for modern steam turbines. The recommended requirements for steam turbines are given in Table 7.6 [reference (7)].

As stated previously, steam purity will vary when operating other than at the design conditions; Table 7.7 gives some guidance as to the operating periods at which steam purity can deviate without serious effect on steam turbine performance and availability. When purity

Table 7.6 Steam purity limits

Parameter	Unit	Guidance value	Ideal value
Conductivity at 25 °C (see `Note' below)	μS/cm	< 0.2	0.1
Silicic acid (SiO_2)	mg/kg	< 0.020	0.005
Total iron (Fe)	mg/kg	< 0.020	0.005
Total copper (Cu) to be monitored if copper materials are used in steam circuit, e.g. condenser tubes	mg/kg	< 0.003	0.001
Sodium (Na)	mg/kg	< 0.010	0.002
For pressures above 100 bar		< 0.05	
For pressures above 165 bar		< 0.03	

Table 7.7 Time limits for off-specification operation

Parameter	Unit	Level 1	Level 2	Level 3	Level 4
Conductivity at 25 °C (see `Note' below)	μS/cm	> 0.2 < 0.35	> 0.35 < 0.5	> 0.5 < 1.0	1.0
Silicic acid (SiO_2)	mg/kg	> 0.020 < 0.030	> 0.030 < 0.040	> 0.040 < 0.050	0.05
Total iron (Fe)	mg/kg	> 0.020 < 0.030	> 0.030 < 0.040	> 0.040 < 0.050	0.05
Total copper (Cu) if applicable	mg/kg	> 0.003 < 0.005	> 0.005 < 0.008	> 0.008 < 0.010	0.01
Sodium (Na)	mg/kg	> 0.010 < 0.015	> 0.015 < 0.020	> 0.020 < 0.025	0.025
Maximum continuous operation (hours)	h	< 100	< 24	< 4	0
Maximum accumulated hours per annum	h/a	< 2000	< 500	< 80	0

values deviate from those given in Table 7.6, in order to avoid permanent damage to the turbine action needs to be taken to restore them within the time limits given. Start-up is a typical condition that can result in level 2 limits, which should continuously improve as the plant gets under load.

Note

The conductivity levels given in Tables 7.6 and 7.7 are to be measured downstream of a strongly acid cation exchanger with continuous measurement at the sample point. Measurements should be taken both to monitor the steam purity and the purity of the returning condensate. Conductivity measurements provide an indication of the presence of salts and give warning of excessive dosing of water treatment chemicals and, perhaps more importantly, the leakage of raw water into the condensate as a result of tube leakage. They do not replace the need for regular chemical analysis to monitor the other contaminants listed. The presence of oxygen has to be avoided by the use of degassing techniques and any residuals removed by chemical dosing. Hydrazine, which is commonly used to remove oxygen, breaks down into ammonia causing the corrosion of the copper alloys used for condenser tubing. Monitoring of copper contamination gives warning of this. Excessive corrosion will lead to cooling-water leakage.

Steam conditions

Steam turbines can be made in a range of materials and the materials selected must match the specified operating conditions. Operating with any abnormal excess temperature steam condition for however short a period will affect its reliability. It is therefore important to nominate the most severe conditions of operation for material selection for the turbine.

Condensing turbines are designed for specific steam inlet conditions. There is a design limit to the moisture content that can be tolerated within the turbine that should not be exceeded. If steam is supplied at lower pressure and/or temperature this will result in an increased amount of water being condensed in the low-pressure section. Operating with cooling water at lower than design temperatures will produce a greater vacuum and have the same effect. It is important to ensure that all possible operating conditions that can affect the steam supply conditions or vacuum conditions are fully specified in the design for the turbine. In some cases it may be necessary to ensure that operating parameters are kept within limits acceptable to the turbine. Abnormal amounts of water entrained with the steam will cause blade damage. Special design features are needed to allow a greater degree of moisture, and facilities to separate and drain water into collector belts are needed. When provided, these drainage facilities are a critical maintenance item; should they become blocked erosion damage to the blades will occur.

Chemical contamination

In some steam circuits the need to by-pass steam at some stage of operation may be required. This involves a pressure drop and the need to use water injection to control the temperature. The composition of this water will need to be checked for any adverse corrosive effects.

Water slugs

Any steam distribution system with pockets or dead legs is also likely to accumulate water. Water slugs if entrained with the steam supply to a turbine can cause severe damage. Back-pressure turbines on standby are prone to have steam condensing in the turbine casing. Start-up without adequate blow down will also result in damage.

Low-load operation

Multi-stage turbines can overheat at low-load operation. This is due to reduced steam flow over the last rows of blades with frictional heat due to vortex effects and windage. If the required mode of operation causes excessive overheating then cooling-water spray will be needed. This gives rise to the danger of excessive water injection, which must be prevented as it will result in erosion of the trailing edge of the blades.

Stopping and starting

Shutdown of condensing turbines for any extended period under vacuum can result in corrosion damage due to air leakage into the casing. The requirements needed to avoid this will depend on the length of time the turbine is shut down. Procedures will range from doing nothing for short and infrequent periods, to, in the event of longer periods of shutdown, bleeding in steam to prevent air entry, or purging with dry air or nitrogen; in the extreme case mothballing may be required.

Any turbine on shutdown will cool and accumulate condensate in the casing. It may also be necessary to slow roll or bar over the rotor so as to prevent thermal distortion (sagging). On re-starting a warm-up procedure with a small steam flow through the turbine casing is required to heat the rotor and casing. Any condensate in the casing has to be discharged to prevent damage due to water entrainment. During this period the steam flow will need to be increased by a small amount, just sufficient to slow roll the rotor. This must then be followed by a rapid acceleration and avoidance of any critical speeds up to the minimum continuous speed ready for load. These procedures are important and need to be reviewed with the turbine vendor. Failure to carry out these procedures may lead to excessive temperature differentials across the rotor inducing creep and thermal fatigue failure. Design features should be provided to prevent operator mal-operation. Very often they are overlooked in the piping design so that operators try to make do using the main throttle valve. This can result in start-ups that cause damage to the rotor and labyrinths. Small by-pass valves for manual operation should be fitted for reliable control of steam for warm up and slow roll operation. The means for reliable blow down, steam inerting, and other such requirements will need to be provided. A HAZOP analysis of the turbine installation design should be carried out.

Design hazards

Manufacturers generally have modular designs and so long as any required turbine is within proven experience a reliable design should be available. The risk is with the prototype of such a machine, i.e. one with a rotor span and weight of new design, with possible rotor dynamic problems. Special vigilance during design and testing is needed. It will be especially true of pass-out machines with their more complex rotor arrangements.

Multi-stage turbines due to their many rows of blades and nozzles are designed to avoid any possible blade excitation. When used for mechanical drives the possible excitation from other components, such as the use of gearing and the impulse from the tooth passing frequency, will need to be verified.

Improving reliability
The need to match loads

Mechanical drive steam turbines need to be matched to the driven process machine load. Usually attention is given to the rated power output required but other part-loads are just as important, especially those conditions that are required for operation for extended periods. The valve sizes of the turbine steam throttle need to be checked to ensure that they are sized correctly. They must be sized to match all required loads so as to avoid prolonged operation with cracked open valves with the danger of wire drawing.

Lubrication
Due to steam leakage from the turbine shaft sealing system, there is a danger of steam leakage into the bearing housings. This results in the steam condensing and mixing with the lubricating oil and contaminating the lube oil system. While measures may be used to remove the water in the system, the oil still needs to be checked regularly. The presence of water also gives rise to the danger of micro-organisms being developed. To avoid this, the provision of a dedicated lube oil purifier should be considered for each critical service steam turbine train.

Hazards to health and safety
These will be the hot surfaces of the machine, moving parts, and the danger from any release of steam. The piping systems around the turbine will come under the PSSRs. There is also a possible danger from broken blades but the risk of them coming through the casing is remote.

7.4.3 General-purpose steam turbines

General comments
These are mostly single-stage units with a few rows of impulse blading. They are very useful as mechanical drives operating with a moderate back-pressure into steam mains that supply process heating. As with other types of turbine they must be matched to all the load requirements. For part-loads a number of hand throttle valves are usually supplied mounted in the steam chest. These are used to adjust the turbine nozzle area so that the governor valve operates at the optimum opening for all required operating duties.

Improving reliability
General-purpose steam turbines can be fitted with shaft labyrinth or carbon ring steam seals. Labyrinths leak steam. Carbon ring seals wear and unless maintained will leak steam. Steam leakage is wasteful and causes problems. It affects the linkage of mechanical governors and their reliable operation. Electronic governors will avoid this problem. Steam condensing into bearing housings affects lubrication. Steps for a more positive sealing of bearing housings, such as the use of air mist lubrication or other such measures to provide a positive pressure in the bearing housing, should be considered. Another approach is to consider the use of dry gas seals that are available for shaft sizes up to 150 mm diameter and operating steam conditions up

to 27 bar, 400 °C. Their main advantage is to reduce steam loss while at the same time preventing steam leakage into the bearing housings.

Hazards to health and safety
These will be the same as those for special-purpose steam turbines.

7.4.4 Gas turbines
Gas turbines take in air, which needs to be compressed so that energy is further added by the combustion of fuel for subsequent energy recovery by expansion through a turbine. This energy cycle requires the processing of two streams of two fluids: air and fuel (liquid or gas).

Air process hazards
Air is used for the constituent gases oxygen and nitrogen. Air, however, can contain pollutants that are a hazard. These can cause corrosion, erosion, and fouling of the compressor blades. Adequate filtration of the air is essential for reliable operation. It is a mistake to believe that the use of compressor washing or cleaning facilities will overcome this. On-line cleaning has limited effect and may need to be carried out once a week. Off-line soaking and slow roll cleaning may need to be done monthly. This can result in down time for many hours due to the period of cool down required. These frequencies are perhaps extreme and will depend on the air pollution at each turbine location. With high pressure ratio machines the fouling problem is worse because of the higher temperature at the later stages of the compression cycle. This means that dust deposits become baked on to the blades and are more difficult to remove. Increasing the efficiency of air filtration will improve turbine availability, as it will increase the operating time periods between the need for cleaning. The problems of air filtration depend on the location, and are:

- icing and blocking of the air intake due to transient sub-zero temperatures;
- sand storms in desert locations;
- snow storms in sub-zero locations;
- in farming areas, seasonal pollution from fertilizer dust;
- industrial pollution, especially contamination from oil vapours and corrosive vapours;
- insect swarms in tropical area;
- in marine environments, the need to filter salt from the air;
- the need for filter maintenance;
- the need to resist atmospheric corrosion.

All these considerations have to be taken into account in the selection and design of the air filter system and the inlet orientation. It has been reported that poor filter design can lead to plugging and excess pressure drop resulting in a deterioration of gas turbine output. The importance of investing in efficient and reliable air filtration is often overlooked and requires detailed attention.

Turbine degradation

Any salts and acid pollution of the air will lead to corrosion of the air compressor. Pitting occurs and blades can then fail as a result of fatigue. The combustion system and the turbine section of the gas turbine have to contend with hot gases and the products of combustion. These components suffer from high-temperature oxidation and thermal fatigue. High temperature stimulates the oxidation of the metal atoms and at temperatures above 899 °C rapid metal depletion will occur unless a protective coating has been applied. These problems are made worse by air pollution and fuel contamination. A rapid form of attack called 'hot corrosion' is associated with alkali metal contaminants such as sodium and potassium. Only a few parts per million of such contaminants will react with any sulphur in the fuel to cause this type of corrosion. Sodium is contained in sea-water. It can contaminate fuel and can also enter via the air intake in marine environments. The combustion process will produce sodium sulphate if sulphur and sodium are both present. This results in high-temperature corrosion at temperatures in the range of 816–927 °C. When there is also a significant presence of sulphur dioxide, low-temperature corrosion will occur in the range of 593–760 °C.

Gas turbine reliability is affected by all the above failure mechanisms that will reduce turbine output and efficiency due to loss of clearances, loss of blade profiles, and combustion chamber defects. The rate of degradation is affected by the amount of contaminants in the air and fuel. It is also affected by the number of starts due to the effect on thermal fatigue and the wear on labyrinths that will inevitably be caused. It is usual to expect some 4 per cent loss in output in the first year of operation and some 7 per cent between major overhauls.

A mean time between required major maintenance is typically 25 000 hours. This is affected by:

- firing temperature;
- the type of fuel (gas has less contaminant than fuel oil);

- the metallurgy adopted and the effectiveness of any protective coatings;
- the frequency of stops and starts.

Operation on heavy oil with a large ash content can result in the need for major maintenance work after only 10 000 hours of operation [reference (8)].

Gas turbine power output

Gas turbine power output is a function of its compression ratio and the mass flow of its compressor. Ambient temperature, pressure (altitude), and humidity affect output. Output is usually given for the International Standards Organization (ISO) rating at 15 °C and 1.013 bar (sea level) and 0.0064 specific humidity. The output is reduced by changes in ambient temperature, exhaust back-pressure, and inlet pressure drop. Some ±20 per cent change in output can occur at sites that experience large extremes in temperature.

It is a mistake to design plants that depend on gas turbines to be based on the average annual temperature and expect constant output. This results in operators running at the maximum allowable firing temperature for long periods, which will accelerate blade failure. Ideally these plants should be designed able to operate with varying output so as to match the ambient effects on gas turbine power.

Gas turbine types

Gas turbines are either of the industrial type or aero-derivative type. In the aero-derivative type, use is made of an aero-engine to generate the hot gas. The gas is then expanded through a power recovery turbine for connection to the load. The advantage is its maintainability. The aero-engine part can usually be replaced in one 8-hour shift for maintenance elsewhere. It might be thought that a proven aero-engine will be more reliable. However, a long period of development to prove reliability is still needed. This is because operation at ground level is completely different from operation at high altitude.

Industrial types are designed as a unit from the beginning. They are usually single shaft for constant-speed operation such as for power generation. Two-shaft designs are used for mechanical drives to allow variable-speed operation.

Hazards to health and safety

The noise emission will exceed permitted limits. Regulations on exhaust emissions may preclude some fuels depending on their composition. The combustion process for any fuel will generate NO_x although the quantity may differ. Measures to control NO_x will be needed and may well be a standard feature for some machines. However, it should be noted that dry NO_x reduction might not be effective for loads below 60 per cent. Wet NO_x control, using steam or water injection, is usually effective for loads down to 20 per cent. Part-load operation should always be checked for regulatory compliance.

Other hazards will be the risk from hot surfaces, moving parts, and fuel. Any release of fuel around the combustion area is likely to ignite and cause a fire. Noise attenuation and fire protection are major concerns for any gas turbine installation. If gas fuel is used then gas detection will also be needed to detect any possible leakage.

7.4.5 Synchronous motors

The use of synchronous motors is attractive for high power output because of the facility for power factor correction of the electrical system. They also present a hazard to the machine train of which they form a part. Each time that they are started, they generate torsional oscillations as they increase in speed from standstill to operating speed. At operating speed the oscillations cease when they become synchronized to the power supply.

The torsional oscillations change with speed so that usually at some point the natural torsional frequency of the train will be stimulated. The torsional stress and the number of cycles experienced for each start will need to be evaluated to ensure that the fatigue life of the train will meet the required life of the plant. Should this not be the case then a torsional damping coupling will need to be used to change the natural frequency of the system.

7.4.6 Diesel engines

Diesel engines are commonly used for emergency power supplies for process plant. Usually an estimated maximum load for the essential power supply is used for the sizing of the engine. Some of these loads will depend on circumstances. It is also important to estimate the minimum load required to ensure that this would be acceptable to the engine. Diesel engines, besides having a maximum rated load for continuous operation, also have a minimum load. This is applicable for diesel generator sets where constant-speed operation is required. Operation at the minimum possible load will cause smoke and carbon deposits in the engine and is usually time limited. The minimum load for continuous operation will be at a higher rating. Emergency load schedules may need to be adjusted to ensure that the engine is always operated within its permissible load limits.

Hazards to health and safety

There will be similar concerns with regard to hot surfaces, moving parts, noise, and exhaust emissions as for gas turbines. Diesel engines will have added risks possibly due to the proximity of fuel tanks. Due to engine vibration there is also a greater danger of fractured fuel pipes. If a heated heavy fuel is used, any fracture will release a highly flammable vapour. To avoid this danger sheaved fuel pipes can be used so that any leakage is detected and contained. There is also a danger of crankcase explosions from the ignition of oily vapours. Safeguards are the use of oil mist detectors and explosion blow-out doors with flame traps.

7.5 Summary

Although the above does not list all the types of machinery in use in the process industries, it has been shown that reliability is as much affected by process conditions as it is by the machine. Designing a machine for a specified duty is not enough. All aspects of its application must be explored. Any circumstance that could lead to machine failure needs to be identified. Any risk to health and safety and the life-cycle cost must then be assessed for possible design action.

7.6 References

(1) **Bloch, H. P.** and **Geituer, F. K.** (1999) Proper use of equipment failure statistics. *Hydrocarbon Processing*, January.

(2) **API Standard 614**. *Lubrication, Shaft-Sealing, and Control-Oil Systems for Special Purpose Applications.*

(3) **Marsh, G.** (1999) *Dry Gas Seal System Reliability – A Seal Manufacturer's View Point.* IMechE paper S620/014/99.

(4) **Lelgemann, K. D.** *Process – Gas Screw Compressors for Various Applications.* MAN Turbomashinen AG GHH BORSIG, Oberhausen, Germany.

(5) **Boer, F. L.** *Reciprocating Compressor Piston Rod, a Hidden Design Challenge.* IMechE paper C449/019/93.

(6) **Wilson, R. S.** *Advances in Piston and Packing Materials for Oil Free Compressors.* Compressor Products International, Hungerford, UK. IMechE paper C542/012/99.

(7) **Siemens AG** *T6300 Steam Quality*. Erlangen, Germany.

(8) **Schilke, P. W.** (1996) *Advanced Gas Turbine Materials and Coatings*. GE Power Systems Schenectady, NY, GER = 3569F.

Chapter 8

Keep it Safe

8.1 Introduction

While machines can be designed to be as reliable as possible, their continued safety in operation depends on their control systems. To ensure safety the machine operating parameters need to be controlled and monitored. Process machines are used to handle many types of hazardous gases and liquids and safety relies on the integrity of their containment in pipes and pressure vessels. Furthermore, where the containment of high-pressure gases is involved there is also the danger of explosion, either due to the failure of safety devices or due to the degradation of materials. When the failure of instruments or fluid-containing components gives rise to a danger to EH&S then the maintenance of these items is safety critical. In order to keep process machinery safe it will be necessary to identify and attend to all safety-critical maintenance requirements.

8.2 Control and monitoring systems

8.2.1 Instrument reliability

Machinery needs control systems and instruments to provide continuous surveillance of machine running conditions and the regulation of utilities such as lube systems and cooling systems. In the event of any malfunction, machines are protected by alarms and shutdown signals. The sensing devices that provide these signals lay dormant until danger arises.

As explained in Section 4.4.7, any device can fail at any time and so these sensing devices may not be available to function when required. They need to be tested regularly to check that they are still working. Testing needs to be far less than the MTTBF. The shorter the test interval the more likely it is that the failed device is found and repaired before it is needed to operate. So the probable failure rate is a function of the test interval. Testing every 30 days is usual, but in

Table 8.1   The effect of test intervals on reliability

	Failure rate (f/hour) at test intervals:			
Equipment	Continuous use	10 000 h	1000 h	500 h
Gas detector	0.000 047	0.47	0.047	0.0235
Pressure switch	0.000 005	0.05	0.005	0.0025
Level switch	0.000 03	0.3	0.03	0.015
Temperature switch	0.000 023	0.23	0.023	0.0115
PRV	0.000 03	0.3	0.03	0.015
Pump	0.000 5	5.0	0.5	0.25
Pressure control	0.000 012			
Tolerable λ	0.000 004			

some cases testing every 2 weeks may be necessary in safety-critical situations. Table 8.1 shows the effect of testing intervals on failure rate. It can be seen that the failure rate of any dormant device is high when compared to it being in continuous use. Reducing the period between testing gives some improvement, but not enough. To improve reliability additional design measures are needed.

For example, a pressure control system backed up with a high-pressure shutdown switch together with a pressure relief valve (PRV) has diversity and three levels of redundancy. It is a parallel system and all three must fail before there is a possible explosion. The safety system reliability depends on adequate testing of the pressure switch and the PRV.

In many cases, systems such as fire detection, lube oil pressure and temperature monitoring systems, all rely on single devices. Sensing devices such as pressure switches can fail in a number of ways.

- Setting point drift: set at a higher than required signal so that it will not respond as required. Alternatively it may fail to work for other reasons.
- Setting point drift: set at a lower than required signal so that it actuates when it should not. This is the reason for false alarms and trips.

To improve reliability, fail-safe systems can be adopted but while this will improve safety, it is at the expense of unreliability due to the many false shutdowns that will occur.

For example, if, instead of 1001 (see Table 8.2), only one instrument, 1003 (any one of three) were installed then the system would be more reliable because all three instruments will need to fail for there to be no signal. With redundancy the system will have a much lower failure rate.

Unfortunately with more instruments the number of false trips will be much greater. This can be overcome by the use of voting systems. This is illustrated in Table 8.2 [reference (1)], which compares the performance of systems with different degrees of redundancy and different voting systems.

By comparing the number of false trips and the number of times the systems fail to work, it can be seen that a voting system of two out of three is the best compromise. This is the system usually adopted for safety-critical signals. In the table, the failure rate λ of 0.5/year is the same for false signals as that for failing to work. T, the interval between inspections, is 0.25 years. Another advantage of voting systems will be the facility for on-line maintenance that is available due to the installed redundancy. The system will be at greater risk during the maintenance operation and special maintenance procedures will be needed to offset the risk.

Another approach is to avoid the use of all switches. All parameters are monitored with transmitters that are in continuous operation and monitored, so that alarms and shutdown signals are determined by solid-state logic. The transmitters, of course, need regular calibration. In some critical applications dual transmitters are used to provide redundancy. This also allows the on-line removal and

Table 8.2   Performance comparison of different instrument systems

	False trips		Fail to work		Dead time (not working)	
System	Function	No. in 10 years	Function	No. in 10 years	Function	% time
1001	λ	5	λ	5	$0.5^*\lambda\,T$	6.3
1002	2λ	10	$\lambda^2\,T$	0.63	$0.33^*\lambda^2\,T^2$	0.5
2002	$\lambda^2\,T$	0.63	2λ	10	$\lambda\,T$	12.5
1003	3λ	150	$\lambda^3\,T^2$	0.078	$0.25^*\lambda^3\,T^3$	0.05
2003	$3\,\lambda^2\,T$	1.875	$3\lambda^2\,T$	1.875	$\lambda^2\,T^2$	1.56

replacement of transmitters for testing and maintenance. The use of two transmitters will also allow the trending of the signal deviation of the two signals and the use of statistical analysis to forecast the need for maintenance. The whole control system will then finally depend on the reliability of the electronic hardware. Fortunately the reliability for such hardware has been under development for decades. The reliability of these systems depends on the working environment. Under ideal conditions, with the use of redundant printed card circuits coupled with self-diagnostic alarms, these systems should provide 3 years uninterrupted operation.

8.2.2 Maintenance provisions

Plant and machinery need to be designed for instrument maintainability. Facilities to replace instruments while on-line are needed with valves to enable their isolation and venting. To avoid the dead time, facilities can be provided to install and connect replacement instruments in parallel before removing the one that is to be replaced.

In some cases the regular testing of a complete emergency shutdown (ESD) circuit is required. This then needs facilities to isolate the output signal from the shutdown device. The hazard is if a genuine ESD should arise while this is being done, or that the output signal isolation is not removed after test, or that unauthorized isolation is carried out. Any maintenance operation that temporarily inhibits a signal is a hazard to safety and must be safeguarded against with engineering safeguards and maintenance procedures to avoid the risk.

8.2.3 Software development

The use of software and computers is the automation of human manual activities. For the average process machine the control logic required is relatively simple and normal acceptance procedures will verify its correct functioning. For more complex plant such as a steam turbine driven reciprocating compressor with unique features, software will need to be specially designed for the purpose. To believe that the automation of these processes can be taken care of by the computer engineers is a huge mistake. The problem is that while the software engineer has to understand the needs of the user, his elicited knowledge can never match the knowledge and experience of the user. There will always be a communication gap. Measures need to be adopted to ensure that the software is safe and reliable. Examples of these methods are given below.

Assignment of an engineering team

This should be made up of engineers from operations, maintenance, process design, and safety, and a rotating machinery specialist. This team then needs to be matched by a design team from the machinery supplier consisting of their design engineer and software engineers from their software subcontractor.

The management of software development

The recommendations of ISO 9000 *Quality Management and Quality Assurance Standards, Part 3: Guidelines for the Application of ISO 9001 to the Development, Supply, and Maintenance of software* should be followed. These include the need for:

- Adequate specifications from the purchaser and facilities for mutual collaboration. See Table 8.3 for software documentation.
- Establishment of the security of access required by the purchaser; for example, for operational access, for engineering to adjust settings and access stored data, and for management of access to production data.
- The provision of facilities for maintenance and future modifications as and when needed.
- The planning of the work and the need for phased development and quality control (QC).
- Design and implementation of the work and the QC of design activities for the required inputs and outputs.
- Phased acceptance testing and QC.
- Final factory testing of part-assembly of hardware interlinked components to the assembled computer hardware components, to verify communication protocol.
- The use of injected signals to verify cross-communication and logic function.

The work to develop the documents listed in Table 8.3 requires a dedicated effort and may run into many revisions before completion. The work should ensure that problems are found and resolved during the software development so that operational problems in the field are avoided.

Table 8.3   Software control documentation

Document	Purpose	Originator
Operating philosophy	Describes the required modes of operation, control functions, control hierarchy interfaces system, and hardware architecture	Client operations
Specification	Deals with the hardware architecture in specific detail in relation to the selected hardware supplier	Purchaser's engineering team in collaboration with hardware supplier
Process system studies	To conduct a HAZOP of the machine/process system. To identify all machine failure modes, ESD functions, and all modes of operation (start, load, stop, etc.)	Purchaser's engineering team in collaboration with the machine supplier
Design basis	Specifies the detailed requirements for each control function. Identifying the conditional parameters for execution, where the required data comes from, and to which items commands are to be directed for execution	Ditto
Functional design specification	Details the software functions including all required algorithms, input/output signals, and the software phases (modules)	Purchaser's engineering team in collaboration with machine supplier and software engineers
Software and display factory test procedure	Details the test set-up and test procedure for each module and the final factory test mock-up of the system to test sample interface signals	Ditto

8.3   Pressure systems

8.3.1   Introduction

Process machinery packages, especially reciprocating compressors, have process piping, heat exchangers, and pressure vessels. Depending on the type of machine, process fluids can be liquid or gaseous. Pressure systems of pipes and vessels work silently and unnoticed until they fail. Failure can be a hazard to EH&S. The result of failure, depending on the process fluid, could be:

- a sudden emission of a high-pressure jet stream;
- leakage of flammable fluids;
- leakage of toxic fluids;
- leakage of liquefied gases that flash into vapour;
- leakage of extremely hot or cold fluids;
- an explosive rupture;
- whiplash of piping due to attachment failure.

All these events represent a danger to the safety and health of people, the environment, and property. To ensure safety it is necessary to recognize the failure modes that could result in these events.

8.3.2 Over-pressure

Over-pressure of pressure systems will cause a rupture and discharge in the case of liquids and an explosion in the case of gas. This can be caused by a control system failure due to an ineffective design and/or inadequate testing and maintenance.

Another cause of over-pressure is a fire. This causes over-pressure for two reasons: heating of the contained fluid causing expansion, and heating of the containment materials, which causes a reduction in its strength. In the event of a fire, if pressure systems that contain gas or liquid gas are not isolated and vented, the pressure system will rupture at a lower pressure than the safety device settings.

A pressure system may be adequately designed for the duty but can be weakened by damage, for example dented or bent in transit. Unauthorized or improperly designed welded attachments or weld repairs can cause a weakness in the structure.

A change in process operations with higher temperatures or failure to maintain piping supports, guides, and constraints can lead to nozzle overloads and subsequent failure. This can also be the result of inadequate design of connecting pipework. Any modifications to existing pipework, unless carried out by qualified design engineers, will have the same risk.

8.3.3 Materials failure

Materials can fail for many reasons. The reasons may be well known but often, due to a range of circumstances, failures still occur and must be safeguarded against. Damage in transit or during storage on site can be significant and should be safeguarded against.

Failure due to temperature

Carbon steels, if exposed to temperatures below freezing, can become brittle. When operating below freezing, small defects can become critical leading to catastrophic failure unless low-temperature ductile steel is used. They will then fail at a lower pressure than design and lower than the set pressure of protective systems. The need for low-temperature steel can be overlooked where items intended for operation in the tropics then need to transit through sub-zero conditions. Joule–Thompson effects during blow down can drop temperatures below zero. Equipment normally operating in heated buildings may suffer sub-zero conditions due to an accident of some sort to the building and heating system.

Creep

Creep can be defined as the *time-dependent* component of plastic deformation of a material. For equipment operating at elevated temperatures (typically over $0.4T_m$, where T_m is the melting point, approximately 400 °C for carbon steel) creep damage accumulation can be an issue. Rupture life and creep rate are very sensitive to stress and temperature. Any change in operating conditions if overlooked could lead to early cracks in the material.

Thick materials subjected to a severe temperature gradient between the inside surface and the external surface will be subjected to an additional stress due to differential expansion between the hot side and the cold side. Material degradation will accentuate this and results in thermal cracking.

Creep cavitation occurs in areas of high stress concentration under creep conditions. Dislocations (faults in the atomic lattice) in the microstructure will tend to migrate to the grain boundaries causing voids at these boundaries. These voids will coalesce eventually giving rise to cracks.

Thermal fatigue

Pressure systems that are subjected to temperature cycles can also suffer thermal fatigue. This will occur if there are any stresses caused by differential expansion. These stresses will change with temperature variations and thermal fatigue can result.

Fatigue

Materials will ultimately fail due to cyclic stress. A pressure system that operates with a cyclic change in pressure could fail due to fatigue. A change in plant operations that changes the cycle of operation or is started and stopped more frequently could be reducing the as-designed service life. Failure could become more imminent.

It should be noted that fatigue limit data for every type of steel are not always available. However, data in the form of stress–frequency (SN) curves are available for commonly used pressure system steels. These show the relationship between cyclic stress and frequency. The cyclic stress that will cause failure depends on the number of cycles. The number of cycles to failure will be higher for low cyclic stress and lower for high cyclic stress. These data are obtained from testing specimens and so there is a degree of uncertainty involved. The value to be used has to be adjusted by a number of factors based on:

- the thickness of the materials used.
- the presence of any stress concentration.
- corrosion;
- heat treatment;
- notch sensitivity;
- residual stress;
- temperature;
- deviation due to tolerances in the material composition.

Design integrity has to stem from gaining experience from using the same material on similar applications over many decades. Even so, manufacturing tolerances and variations in the quality control of fabrication will affect fatigue life.

The onset of fatigue failure is usually indicated by the initiation of a tiny crack in the areas of the highest stress. At first the crack grows slowly, and then escalates rapidly until fracture occurs.

Machinery or flow-induced vibration can occur as a result of turbulence from the operation of valves. Induced vibration from the main pipework will very often result in fatigue failure of attachments, such as drain and vent connections and instrument lines. Their possible vibration is usually overlooked during design and even if considered might be difficult to define. To avoid failure they should be surveyed during initial operation and vibration data obtained by the use of friction-type strain gauges. These data will then allow analysis to determine if there is any danger of fracture and the need for remedial action.

Fluid flow induced failure

Erosion and erosion-corrosion is caused by the velocity of fluids across the metal surface. This can be due to the abrasive effect of hard particles hitting the surface and can also be combined with corrosion attack as a result of exposing the metal surface of any oxide film. This is known as flow accelerated corrosion (FAC).

Heat exchangers are designed for turbulent flow, but strong vortices can be generated due to the vena contracta effects at the tube entrance. On sea-water service, depending on the amount of entrained solids, the turbulence can result in tube failure. This is a common problem in coastal waters and the use of nylon inserts about 10 diameters long to protect the inlets of the tubes can prevent tube failure.

Cavitation is another form of corrosive attack caused by the formation and collapse of vapour bubbles impacting on metal surfaces. This occurs as a result of hydraulic effects in the operation of pumps, hydraulic turbines, and propellers, etc. and is well known to mechanical engineers.

Fluid velocity also has a great effect on the corrosion rate of materials. There is a critical velocity at which the corrosion rate will increase rapidly. This will differ for different materials and different environments.

Material defects

This can result from materials manipulation and fabrication processes. The inclusion of materials defects and impurities causes local hardness and other deviations of physical properties. The welding processes in fabrication will affect the physical properties of the material in the area of the weld. These problems are well known and can be avoided by the proper selection of weld procedures and subsequent heat treatment. Materials defects can be found by inspection techniques. These all depend on quality control, which is never perfect. Any defective areas missed are then often the source of corrosion. Corrosion can also result from any variations in material composition.

8.3.4 Corrosion

Failure due to corrosion is probably the most common cause of pressure system failure. In theory, by adherence to standards as issued by the National Association of Corrosion Engineers (NACE), corrosion can be avoided. The International Standards Organization (ISO) have published some 64 standards alone. There is no lack of information on how to control corrosion. The techniques of predicting the corrosion rate and developing materials that can inhibit corrosion should give confidence in the design integrity of pressure systems. However corrosion failure still occurs due to:

- lack of training and education;
- cutting overheads and the loss of expertise;
- hazards from the fabrication processes due to ineffectual QC;
- changes in operating conditions;
- extending the operating life of plants.

Because of the uncertainties listed above it is mandatory to inspect systems regularly to check that they are in a fit condition for further operation. The reliability of these inspections depends on knowing:

- the symptoms;
- where to look;
- how to find defects;
- how to predict the residual life.

Types of corrosion and their symptoms
Galvanic corrosion

Most corrosion is due to galvanic action. Galvanic action is caused by electrolytic action like a battery. There need to be two different metals in electrical contact with each other and they must also be submerged in a conducting liquid in order to form a circuit. One is the anode where the corrosion occurs. The cathode is the metal where no corrosion occurs. Electric current leaves the cathode via the physical contact and returns via the conducting fluid, the electrolyte. The rate of corrosion will depend on the relative areas, the distance apart, the resistivity of the electrolyte, and the chemical composition of the fluid. Corrosion can only take place if there is a potential difference and there is an electrical circuit in place.

Galvanic tables are published that show the electrical potential between different metals. Those at the top of the table compared with those at the bottom will provide the greatest potential. The abbreviated Table 8.4 is given to show the relative position of mill scale and weld scale. The table demonstrates why galvanic corrosion is so common and why mill scale and welding oxide layers are often the cause. It also shows the risk of pitting caused by any local damage to the oxide film of stainless steels.

Table 8.4 Galvanic series

Anode end

Magnesium, aluminium, manganese, etc.
Zinc
Steel or iron
Stainless steels without an oxide film
Lead
Tin
Copper alloys
Oxide films
Mill scale; weld scale; welding oxide layers (due to insufficient inert gas shielding)

Cathode end

Pitting and crevice corrosion
Rapid pitting occurs wherever there is a small area of anode surrounded by a large area of cathode. Pitting is also caused by differences in the metal surface such as:

- impurities;
- grain boundaries;
- local surface damage from nicks;
- rough surfaces.

Metal exposed to air will very soon produce an oxide layer that will protect the surface from further corrosion. In the case of carbon steels, oxide films are usually defective and are not protective. Steel alloys form a strong oxide film but any localized damage to this layer will result in an anode being formed and rapid corrosion pitting will follow if the film is not restored. Another example is weld areas where there is a local defect that is anodic compared with the base metal, such as due to a local depletion of alloying elements.

The presence of chloride ions is a particular threat to stainless steel. They have the power to break through oxide films and cause pitting. This is of particular concern for plants using sea-water cooling. In coastal locations its presence in the atmosphere will be sufficient to corrode stainless steel pipework if it is not painted for protection.

Crevice corrosion is the result of a local change in environment. They are the result of oxygen concentration cells in a stagnant space so that the corrosion is restricted to a very small area in a similar way to pitting. Typical sites are:

- holes;
- gasket surfaces;
- lap joints;
- under surface deposits;
- crevices under bolt heads, etc.

Corrosion occurs under welding oxide layers and also under surface deposits and under bolt heads on stainless steel where there is less exposure to oxygen than the bulk material. This can be sufficient to generate a potential difference and cause corrosion. Tubes of heat exchangers that are not correctly rolled into the tube sheet can have cavities that will cause crevice corrosion. Socket-welded flanges that are not seal welded on the inside will have cavities that could corrode. Flanges with fibrous gaskets that allow liquid to be trapped between their faces can also be a problem.

Velocity effects
In many cases pumps that are in operation will not corrode but corrosion rapidly takes place under stagnant conditions. Stagnant conditions allow corrosion cells to develop and this can be avoided if the pumps were to be drained, flushed, and dried out when on standby.

Microbial corrosion
It is possible that up to a third of all corrosion is caused by micro-organisms and practically no materials are immune from their attack. Micro-organisms consist of bacteria, fungi, and mould. They need heat, humidity, and nutrients to become active and cause destruction. Some need oxygen (aerobic bacteria) and others do not (anaerobic). Nutrients can be organic or inorganic. They adhere to metal surfaces and form a gelatinous film.

Sulphate-reducing bacteria (SRB) predominate in anaerobic biofilms that are associated with sulphur-containing liquids such as sea-water and fuel oil. They reduce sulphate to sulphide, which corrodes most alloys including stainless steel. Fuel oil is converted to sludge and is contaminated with gummy deposits. The sludge lies at the bottom of fuel tanks and causes corrosion. Contaminated fuel oil gums up fuel systems and contributes significantly to diesel-engine down time.

Pressure systems need to be hydro tested as the final QC action before being ready for start-up and commissioning. If the water is contaminated in any way, SRB will start corrosion almost immediately unless the water is drained, and the plant is dried out. In one case, water was left in a condenser for a month and on start-up all the tubes leaked due to the pitting corrosion caused by microbial action. It is common practice to use biocides to kill off the micro-organisms. Very often the residual debris will form deposits on tank bottoms and pipework that are a further cause of corrosion. It is far better to ensure that the accumulation of water is avoided and that any water is removed before damage occurs.

Stress corrosion cracking

It is sometimes thought that pitting corrosion will not lead to a catastrophic failure. In some cases it may be true, for example in pipework under low stress. Corrosion pinholes appear on the surface with seepage of liquid to give warning of deterioration.

Stress corrosion cracking will occur where there is a susceptible microstructure in the material under environmental stress. For pressure systems that have areas of stress concentration, the bottom of the corrosion pit itself becomes a further stress concentration. Due to the loss of load-bearing area as a result of the pit, the stress is increased. Stress is further concentrated at the tip of the pit so that a crack is induced. This is hidden and unseen. The combined effects of the increasing corrosion and the consequent increase in stress then accelerate the propagation of the crack until fracture occurs. Stress corrosion can only be avoided by the selection of resistant materials, correct heat treatment, and the removal of corrosion specifics in the environment. These effects are also applicable to machine components such as pump shafts that are exposed to the corrosive environment.

Hydrogen embrittlement

Atoms of hydrogen can rapidly defuse into steel alloys. This can happen in the processing of hydrogen-rich hydrocarbon gas. In other cases atomic hydrogen can be one of the products of a corrosion reaction with liquids that contain H_2S, HCN, HF. The free hydrogen atom enters the metal before it finds another hydrogen atom to form a molecule. Hydrogen molecules cannot defuse into metal. The hydrogen atoms tend to gravitate into voids and other spaces in the metal to form molecules. If the metal is heated sufficiently the hydrogen dissolves into the metal as atoms and disperses freely in the material. On cooling at the transition temperature, the hydrogen atoms seek open spaces in the material lattice to concentrate and reform into molecules. This is usually at locations where the metal is under greater stress. Each time there is a temperature cycle the hydrogen pocket will be under increased gas pressure and more hydrogen will be concentrated at the space so that a crack will develop. High-strength materials are particularly susceptible to this problem, which can mostly be avoided by heat treatment and correct selection of material composition.

8.3.5 Corrosion protection

The best protection is investing in expertise in a design team consisting of the process, mechanical design, and materials engineers. By applying expertise early in the design stage most problems can be avoided by the proper selection of materials and design. One measure to protect equipment is the application of a protective coating to form a barrier between the environment and the metal. These coatings can range from an oil coating to metal plating. The problem of coatings is a technology in itself. Will they be affective and for how long? Badly applied coatings with pinholes can accelerate corrosion. Any penetration of the coating, such as by a drilled hole that causes exposure of the base material, can be a site for concentrated attack. Other measures involve changing the direction of current flow to prevent corrosion. This can be by use of a sacrificial anode such as zinc or by the imposition of direct-current impressed current cathodic protection. Care has to be taken in designing and using impressed currents since too high a current can lead to hydrogen generation and embrittlement of the component being protected. Other measures involve the use of inhibitors and water treatment. These all have their problems and need expertise in their application and maintenance.

8.4 Keeping it safe

8.4.1 Introduction

The foregoing is intended to underline the fact that pressure systems are inherently hazardous. Besides the need to ensure that their control systems are well designed and maintained, pressure systems have a life limitation. The data on which they are designed are never perfect and so their life cycle cannot be predicted with certainty. To ensure safety a regular inspection program is needed to find any onset of damage and to assess the rate of damage thereafter so that the equipment can be repaired or replaced before any catastrophic failure. The Safety Assessment Federation (SAFed) recommended inspection intervals are given in Table 8.5.

Table 8.5   Inspection intervals

Pressure system type	Frequency (months)	Notes
Air pressure plant	26	
	48	For well-maintained plants of welded construction
Hot-water boiler (operating at 100 °C and over)	14	
Refrigeration and air conditioning	26	For systems over 25 kW
Steam boiler and steam oven	14	
Steam pressure vessel	26	
Other pressure systems	26	

Inspections are statutory requirements. The frequencies shown are recommendations. They must be adjusted based on actual usage and risk assessment for each situation.

8.4.2   Failure statistics

The importance of in-service inspection is underlined by the compilation of failure statistics of actual inspections that were carried out over a period of time. The distribution of failures drawn up from the results of inspections carried out on plant in an industrial area of the UK is shown in Table 8.6 [reference (2)]. The probable causes of these defects are given in Table 8.7. These are preliminary results as compiled by the HSE from data supplied by SAFed and reported in 2002 [reference (3)].

Table 8.6   Distribution of defects found

Type of failure	Percentage found
Corrosion	34
Stress corrosion cracking	22
Fatigue	14
Welding faults	8
Erosion	6
Brittle fracture	3.5
Mechanical failure	3.5
Creep	2.5
Overheating	2
Over-pressure	2
Other	2.5

Table 8.7   The percentage distribution of root causes of defects

Root cause of defect	Heat exchangers	Piping	Pressure vessels
Operator error	5	1	1
Improper design/construction	2	4	4
Improper installation	7	11	10
Poor maintenance	11	17	15
Control/protective device malfunction	15	22	45
In-service defect	60	45	25

For pipework, the causes of pipework failure ranked in descending order have been reported.

- Leakage at flanged joints.
- Leakage from corroded pipe (especially under lagging).
- Leakage at small-bore piping (e.g. due to fatigue).
- Failure of supports.
- Leakage at bellows.
- Leakage at instruments.
- Failure of steam trapping.
- Modifications.
- Wrong materials.
- Over-pressure.

This list correlates quite well with those for vessels as given in Table 8.7.

The statistics in Tables 8.6 and 8.7 show that in spite of regulations, codes of practice, and QC/QA procedures, mistakes still occur and it is the responsibility of inspecting engineers to root them out before a disaster happens.

8.4.3 Risk assessment

The risk assessment must first classify the system with regard to the consequences of failure. The second step will be an audit to verify its design and manufacture. Finally it will be necessary to determine the probable safe operating life. The probable safe operating life will depend on the reliability of:

- the design process;
- the materials physical properties data;

- QC and QA of the manufacturing process;
- operating conditions;
- operating environment;
- instrumentation and control devices;
- the predicted life cycle based on the fatigue life, corrosion rate, etc.

The risk assessment should be carried out prior to operation. This will establish the base line having verified that the basis of design still matches the operational intent. If all the QC and QA documentation is in order then the initial risk of failure should be low. If there is any risk of corrosion then the measures adopted to avoid early failure must be audited.

For example, if the protective coating adopted for corrosion protection is incorrectly applied, a small defect could cause failure within 1 year. This underlines the importance of experienced inspection and QC/QA.

Subsequent risk assessments should audit any deviations from the baseline condition together with the results of inspection. Any changes in operating conditions or of fluid composition, however small, could have a dramatic effect. The inspections should provide evidence of corrosion rate and any sign of impending failure such as the appearance of cracks. The monitoring and trending of such information can then be used to forecast life expectancy and indicate the required frequency of inspection. The strict application of this procedure is the basis for *risk-based inspection*.

8.4.4 Risk-based inspection (RBI)

RBI is a process for the management of risk. Following on from an initial risk assessment the identified risks can then be adjusted based on the subsequent inspection results. From the many possible failure modes the front-runners can be identified and monitored closely and action taken before failure and possible danger to life and limb occurs. It should ensure that any changes are identified so that a new risk assessment can be made. In the course of time new failure modes may be identified and become more critical. The objective is to ensure that inspection programs are matched to the risks as they develop or change. This should enable the critical risks to be monitored so that equipment can be repaired and taken out of service before there is a disaster. For the purposes of RBI, risk assessment should have six stages of development.

(1) Identification of the risk to EH&S from equipment failure.
(2) Identification of the various degenerating effects on materials as a result of the operating environment.
(3) Reviewing the equipment and its operating environment for all the possible modes of failure.
(4) Determining the in-service defects that are associated with the modes of failure and how they should be found and monitored.
(5) Determining which failure mode is likely to cause failure of what item of the pressure system, and how the risks from any defect found can be assessed.
(6) Ranking and categorization of risk from each failure mode.

The identification of risk to EH&S of people has been listed at the start of this chapter. The risk assessment process flow diagram is given in Fig. 3.1. The degree of risk will depend on its probability and the consequence. The procedure for this, as determined by use of a *risk classification matrix*, is shown later.

A recommended checklist of failure modes is namely:

• instruments and protective systems;
• corrosion;
• creep;
• fatigue;
• stress corrosion cracking;
• brittle fracture;
• buckling;
• operator error.

A more definitive list of deterioration mechanisms can be found in API 571 *Potential Damage Mechanisms for Refinery Engineering*. This lists all the deterioration mechanisms, manufacturing defects, failure modes, and the circumstances in which they occur.

A risk assessment of the design and process application then has to be carried out. Each of the possible failure modes that are applicable needs to be identified and reviewed. The vessel will have been designed to the required specification. Based on this the equipment life will usually be limited by the area with the highest stress concentration. This needs to be ascertained from the design dossier.

For example, the stress profile of a piping system will usually show that the seat of initial failure will be located at a nozzle.

The review of a vessel design dossier may show that failure will be initiated from the reinforcement for an access manhole.

The skirt of a vessel has to carry the structural loads of a vessel. There may be exceptional wind gusts. There could be thermal fatigue due to temperature differences between the skirt and the vessel body as a result of unexpected process operations. All this can cause failure at the skirt attachment area.

The mode of failure at the seat of failure can be one of many. Normally the various modes of failure cannot be designed to occur simultaneously. Each type of failure will need to be ranked by its expected endurance limit. These will in turn be dependent on the rate of attrition such as by the:

- number of thermal cycles;
- number of pressure cycles;
- corrosion rate;
- changes in fluid composition.

For the initial assessment prior to operation the life expectancy as designed for all failure modes cannot be assumed. A failure mode could be identified that had not been allowed for in the design. A fabrication defect could come to light as demonstrated in the failure statistics given above. The process operating regime may have changed from that envisaged.

Based on the findings a *written scheme of examination* must then be prepared.

8.4.5 A written scheme of examination

For whatever inspection strategy, a written form of examination is a statutory requirement. A competent person is needed to prepare this. The topics that are applicable will depend on the specific equipment but they should include:

- scope of inspection;
- names and tag numbers of all items within scope;
- work permit procedure to enable inspection to take place;
- a plan of inspection;
- NDT techniques to be used;
- specific areas of special concern (location of possible failure);
- audit of inspection records of instrumentation since the last inspection;
- inspection and test of all instruments and controls;
- review of the NDT inspection results;
- list of remedial work required as applicable;

- issue of a report on the completion of inspection;
- QC and QA procedures for the control of remedial works;
- issue of a certificate of Fitness for Further Service as applicable;
- frequency of further inspections to be carried out;
- any amendments to the procedure found to be required;
- the maintenance of a safety dossier with all inspection records and risk assessment reports.

Although the regulations specify a competent person, this probably is only applicable for standard systems in factories. In the case of process plant the competent person should be made up of a team consisting of the process, design, safety, and materials engineers. Ideally it should be the same team that conducted the risk assessment. The process engineer is needed to identify all the possible process variations, the design engineer to identify the high-stress areas of the design and the possible failure modes, aided by the materials engineer. The safety engineer is needed to review HAZOP reports and work permit procedures. Any possible failure as a result of operator error will need to be identified. The provisions to reduce the risk of operator errors must then be reviewed and verified to be in place. The safety file from the plant design must also be examined and updated as necessary.

8.4.6 Inspection requirements

Instruments and controls
As stated above, instruments need to be regularly tested and calibrated. This includes pressure relief valves. In addition they should also be audited. API RP 576 *Pressure Relieving Devices*, second edition gives a list of 14 issues to be checked during an on-line inspection that should be carried out in addition to testing to ensure that the total installation is in working order. These should include:

- checks to see that the vents on valves sealed by bellows are open and clear;
- checks to see that the vents on discharge stacks are open and clear;
- checks to verify that the correct valve is installed;
- verification that the set pressure as marked on the tag is correct for the system;
- checks that vent pipe supports will prevent reaction loads on the valve;

- verification that all associated block valves, etc. are in the correct position and locked accordingly;
- checks for any leakage;
- inspection for signs of corrosion and deposits that could affect operation;
- checks of test QC, QA documentation.

Pressure system inspection methods

These should be based on the likely failure modes to be encountered. Table 8.8 gives the range of methods needed. However, the inspection process must always be alert to the unexpected, which is always likely to arise.

In order to determine residual life and avoid catastrophic failure it is necessary to detect any surface-breaking defects and to measure how far they extend below the surface. The use of standard ultrasonic techniques to measure anything less than 3 mm deep is difficult and inaccurate and more specialist methods are needed.

Time of flight diffraction (TOFD)

This is a specialist ultrasonic technique that can provide a more accurate measurement of the size and depth of a defect. It is an emerging technology and may not be universally available.

Table 8.8   Choice of methods for different failure modes

Failure modes	Cause	Method to use
Internal wall thinning	Internal corrosion Erosion Cavitation Weld corrosion	Ultrasonic thickness measurement Radiography
External wall thinning	External corrosion Corrosion under insulation	Visual inspection Radiography Thermography
Cracking	Fatigue Stress corrosion cracking Wet hydrogen cracking	Ultrasonic Radiography Magnetic particle Liquid penetrant
Other	Creep Hot hydrogen damage High temperature	Ultrasonic Radiography Magnetic particle

Alternating current potential drop (ACPD)
This is an old and standard method for the accurate measurement of crack depth which has fallen into disuse but should not be overlooked.

Eddy current examination by complex plane analysis
This system has now been developed to the point where the ability to detect cracks has reached the same level as with magnetic particle inspection (MPI) but without the need to remove surface protective coatings. Coatings up to a thickness of 2 mm can be tolerated.

Eddy current techniques for inspecting non-magnetic heat exchanger tubes are also available.

Long wave or guided wave technology
This uses the properties of ultrasound inspection for the detection of corrosion under insulation. It is useful for the inspection of insulated pipe for example. This technology is only applicable for ferromagnetic materials.

Investigation procedure
Fatigue-type defects
First any coatings will need to be removed and magnetic particle inspection carried out.

In the as-welded condition, black particles on a white contrast should show cracks down to 5 mm long by 2 mm deep. The use of fluorescent particles and ultraviolet contrast gives a better sensitivity down to 3 mm long by 1 mm deep.

For surface-breaking defects, normal ultrasonic techniques are not effective for cracks less than 3 mm deep and specialist skill is needed. Time of flight diffraction or alternating current potential drop methods should be used.

An alternative that will operate through coatings of 2 mm or less is the use of eddy current techniques for ferric steel welds. It has the same sensitivity as black particle inspection with the advantage of not needing the removal of any coatings.

Early detection of cracks will allow trend monitoring. The rate of crack propagation can be used to estimate the residual life expectancy. Unfortunately the probability of finding cracks of the length of 3 mm is only 75 per cent and of finding those of 5 mm it is only 85 per cent. The chances of finding those even smaller will be very much less. If the operating environment is conducive to cracks then there is always a 15 per cent chance that an undetected crack is present. In these situations it may well be prudent to estimate the residual life for a 5 mm crack and ensure that the prescribed period before inspection is less.

Corrosion or erosion

Pitting is difficult to detect by normal ultrasound. It depends on the shape of the pit. It requires a well-defined bottom to the pit for a good response, such as a lake-type pit. Cone- or pipe-type pits are almost impossible to find and size. In these cases a magnetic flux leakage system will need to be used. The instrument has to be precalibrated using a model of representative pits created on a similar thickness material.

Methods used for on-line inspection

There are methods for the digital measurement of wall thickness; for example, measurements of corrosion under insulation such as thermography, and of in-service ultrasound surveillance such as long range ultrasonics. Acoustic emission monitoring can be used to locate the propagation of cracks. Each of these technologies needs some expertise in their use and in their interpretation.

Competency in NDT techniques

The foregoing is only an introduction to the subject [reference (4) gives further details]. Competency in the operation of NDT equipment and skill in the interpretation of results requires specialist education and training. In the UK, only organizations or certified technicians as accredited by the United Kingdom Accreditation Services (UKAS) should be used. In other European countries accreditation will be by the relevant national bodies such as COFRAC, AENOR, etc.

Residual life assessment

API 579, second edition, 2003. Fitness for Service

This provides guidance on how to quantify the effect of flaws or damage found during inspection of operating equipment so that a decision can be made on its fitness for service, to either run, repair, or replace. The procedures relate to equipment designed to ASME and API international codes, but care will need to be exercised with regard to European codes. It is intended for application in the petrochemical industry and provides procedures to assess the following:

- fracture;
- fatigue;
- thermal fatigue;
- creep;
- metal loss;
- pitting;
- blisters, laminations, gouges, and grooves;
- weld misalignment, out of roundness, bulges, and dents;
- fire damage and local overheating.

Evaluation procedures provided include:

- statistical evaluation of corrosion data;
- the application of remaining strength factors for locally thinned areas;
- comparison charts for the statistical treatment of pitting damage;
- evaluation of residual stress;
- evaluation of stress intensity;
- evaluation of in-service fracture toughness;
- data and equations to estimate crack growth rate;
- evaluation of fatigue life.

The equivalent British standard is BS 7910; 1999 with amendment No 1: *Guide on Methods for Assessing the Acceptability of Flaws in Metallic Structures.*

Risk ranking

In any process plant there will be pressure systems handling a variety of different fluids. These will range from utility systems to complex process systems. Attention needs to be focused on those that pose the highest risks to EH&S and production output. The hierarchy of risk must be:

- explosion due to failure of gas-containing systems;
- release of flammable and toxic fluids.

Risk ranking consists of identifying those pressure systems that pose the highest risk of failure with the worst possible consequences. The risk ranking will need to be based on the following considerations.

Explosion

The severity of the explosion will depend on the stored energy involved. Besides the immediate effect from the explosion there will be the contingent risk from the gas released. What will be the size of the gas cloud? If it is poisonous will it drift to the nearby village? If it is flammable what are the chances that it will ignite? What are the safeguards in place to contain the incident?

Leakage
Failure of pressure systems that contain only liquid is unlikely to result in an explosion as the storage energy is released when a small volume of liquid escapes. However, the discharge of a high-pressure jet of liquid will be a safety hazard. The most dangerous liquids will be those that will change state with a loss of pressure. Any fracture will result in the sustained discharge of a jet of two-phase fluid. The contingent risk from the discharge of liquid will need to be assessed as to the possibility of fire or danger to EH&S.

Fluid properties
Process plants handle a whole range of fluids. They have different properties, for example diesel fuel is flammable but does not ignite as easily as gasoline. Some affect health in different ways and some are more dangerous than others. They should be ranked with regard to the risk to health, fire, and reactivity as discussed in Chapter 3.

Containment
When any loss of containment occurs the critical action needed will be to limit the loss of fluids and to contain any that escape. This is usually considered in the plant design. The emergency shutdown procedures and the facilities for containment of fluids and their disposal will need to be reviewed.

To contain an explosion, blast walls can be provided. Damage to a building can be limited by the use of blow-out panels. Any nearby control rooms can be designed to be blast proof to ensure the safety of the operators and the security of the control systems.

Consequence
Should any failure occur the consequence would be the resulting harm to people, the environment, and the plant. This can be ranked in accordance with the numbers of people that could be killed or injured or the impact on plant operations.

Risk of failure
The pressure systems must each in turn be assessed for the possibility of failure. A probability of failure assessment needs to be carried out for each of the possible failure modes and the probable residual life expectancy determined. The results should be displayed in a tabular

form listing the failure modes surveyed. Any defects found and the life expectancy for each and any action to be taken to reduce the risk of failure should be noted.

The possibility of operator error should have been considered during system design by the use of HAZOP studies. However, these should be reviewed in the light of operating experience. Associated safety procedures should be audited for effectiveness and any design provisions to prevent operator error inspected.

The tabular ranking of the failure modes for each pressure system needs to be updated following each inspection. Due to changing circumstances the residual life for each failure mode may change. They will be like the horsemen of the apocalypse rushing to disaster. Which one will get there first? Can the front-runner be hobbled to slow it down? What must be done to prevent disaster? These are the dilemmas that face the inspecting team. Or conversely, can an extended operating period before the next inspection be justified? Should on-line inspection be prescribed or an interim audit?

Risk assessment

This needs to be done for each vessel and piping system. The first step is to review the table of failure modes for each vessel or system. The failure mode with the highest risk needs to be identified; the highest risk being the shortest time before failure is expected to occur. The consequences of failure can then be assessed based on Table 8.9. This is a qualitative assessment and is usually based on the judgement of the team members.

Having decided on what the consequences of failure will be, the probability of failure will need to be assessed. This can be done using Table 8.10. The probability of failure will depend on the mode of failure being assessed. In the case of fatigue it will depend on the allowable cycles of operation remaining after quantification of the cycles already imposed. In the case of corrosion it will depend on the residual thickness of material and the expected corrosion rate. If cracks have been discovered by inspection, it will depend on the estimated rate of crack propagation and the critical size at which rupture will occur. The evaluation of these situations will give an indication of the residual life of the vessel or piping system.

Table 8.9 also gives guidance for assigning the risk in a new situation with an existing plant. The risk will then depend on the amount of reliable data available at the start of the process. One of the most important activities required will be the need for retro-

Table 8.9   Consequence assessment

Consequence severity level	Definition (any one or more)
5 *Serious*	In-plant fatality; public fatalities; extensive property damage; serious long-term health/environmental damage; 2 or more days extended downtime.
4 *High*	Lost-time injury; public injuries or impact; significant property damage; health/environmental impact exceeding regulation standards; downtime of 1 to 2 days.
3 *Medium*	Minor injury; moderate property damage; minimum short-term health/environmental damage; 4 to 24 h downtime; disruption of product quality.
2 *Low*	No worker injuries; minor property damage; no health/environmental impact; downtime less than 4 h.
1 *Minor*	No worker injuries, property damages or health/environmental impact; recoverable operational problem.

spective engineering to fill up the data gaps. This will be the problem that a team will face for deciding on the viability of extending the life of an old plant.

The results of the risk assessment can then be recorded on a risk consequence matrix as shown in Fig. 8.1. The team will need to judge

Table 8.10   Probability assessment

Likelihood	Definition (as appropriate)
A *Very unlikely*	Full operating history, design and inspection data available. Deterioration rate known and monitored. No significant fatigue cycles sustained. Expected remaining life > 10 years.
B *Unlikely*	Operating history, design and inspection data not fully complete. Deterioration rate estimated with reasonable accuracy. Fatigue cycles sustained < 20%. Expected remaining life 7–10 years.
C *Possible*	Operating history, design and inspection data reasonably complete. Fatigue cycles sustained < 40%. Expected remaining life 5–7 years.
D *Probable*	Operating history, design and inspection data incomplete. Fatigue cycles sustained < 60%. Expected remaining life 3–5 years.
E *Highly Probable*	Operating history, design and inspection data unknown. Fatigue cycles sustained > 60%. Expected life expired.

	Probability				
Consequence	A	B	C	D	E
1					
2					
3					
4					
5					

Fig. 8.1 **Probability consequence matrix**

what degree of risk and consequence is acceptable. In most cases the risks that fall in the shaded area will be unacceptable.

This work needs to be carried out for all the systems and vessels on the plant so that they can be ranked in the order of highest risk to ensure that attention is focused on the most critical items. If action is taken to reduce the risk of failure of these items, such as design modifications to reinforce weakened areas or action to reduce the rate of corrosion, then a reassessment will be needed. A new or revised set of inspection plans for each vessel or system will need to be issued for the next scheduled inspection.

8.4.7 The management of RBI

From the foregoing it can be appreciated that the management of the RBI of a process plant is a job of some magnitude. Much work has been carried out on the development of RBI procedures, as commissioned by the HSE [reference (5)] and API publications are available on the subject [reference (6)].

Software programs are also available for the management of RBI. Typically they will have features such as:

- a database to capture nameplate and design data for all plant items;
- NDT knowledge base;
- library of damage mechanisms;
- library of process fluids with their EH&S rating factors;

- forms for the preparation of inspection plans;
- risk assessments reports;
- tamper-proof filing of inspection reports and the recording of responsible persons;
- records of the location of any defects found;
- analysis of findings with facilities to provide residual life indication (RLI);
- provide a risk matrix of results for each item or system;
- provide a risk profile of the plant;
- maintaining an audit trail of all inspections and findings for each vessel and system;
- enabling equipment to be ranked in accordance with their RLI;
- allowing input of risk mitigation action plans;
- providing an assessment of proposed action plans on the inspection schedule and RLI;
- enabling revised inspection plans to be drawn up focusing on each damage mechanism;
- the issuing of a management report.

8.5   A warning note

The engineering of equipment for pressure systems and the actions needed to 'keep it safe' is not an exact science. As with any item, the engineering of reliability is straightforward for a standard product on which experience has been built up over many decades. In the engineering of process plant this is much more risky, as is evidenced by the failure statistics shown.

It is hoped that this chapter has given some guidance on these matters and, even with expert advice, engineers will heed the well-worn admonishments of:

Murphy's law – if it can go wrong it will.
There are lies, damn lies, and statistics.

Engineers are therefore warned to be cautious, always take the conservative view, round up on factors to show a higher stress. Statistical derived thickness could be worse than it says, so take the worst case and round down. The management will always want to keep the plant running for just a few more weeks to complete a batch. Maybe when the plant was in good shape it was safe. The problem arises when it happens a few times and people become complacent. The danger is:

'*A spiral of descent into poor judgement.*'

This was the verdict on the space shuttle disasters. The engineers blamed the management for not listening to their warnings. The management blamed the engineers for saying 'if that happens, or this happens it will fail'. To say *IF* is not a warning. At the right time, and if they really believe it, engineers need to learn to say: 'it is highly probable for it to happen and people will die'.

8.6 Summary

This and the final chapter have brought together all the issues covered in this book. Laws and regulations in themselves do not ensure safety. In the end it all depends on the education, experience, expertise, and the dedication of the practising engineers. Unfortunately their education has always been focused on the scientific principles that can be defined and quantified. In the twenty-first century engineers are also called upon to cope with the problems of uncertainty. It is hoped that this book will prove to be useful to them in this task. They will need to be more imaginative and lateral in their thinking and the final chapter, offered as an epilogue, will illustrate this.

8.7 References

(1) **Davidson, J.** and **Hunsley, C.** (Eds) (1994) *The Reliability of Mechanical Systems.* Professional Engineering Publishing, ISBN 0. 85298 881 8

(2) **Bainbridge, H.** and **Smith, G.** (2002) *Survey of the Causes and Frequencies of Defects in Pressure Systems Detected from In-service Examinations.* IMechE paper S881/2002.

(3) **Henry, N.** (2002) *Mitigation of Corrosion and Stress Corrosion Deterioration in Chemical Plant.* IMechE paper S881/2002.

(4) **Brown, W.** (2003) *Review of NDT Techniques Applicable to Pressure Systems.* CAN (Offshore) Ltd. Presented at the Plant Integrity Seminar, IMechE, 2003.

(5) **The Health and Safety Executive, UK** (2001) Contract research report 363, *Risk Based Inspection as a Part of Plant Integrity.* www.hse.gov.uk

(6) **American Petroleum Institute** API Recommended Practice, 580 *Risk Based Inspection* and API 581 *Risk Based Inspection, Base Resource Document.* www.api.org

Chapter 9

Epilogue

9.1 Introduction

Many failures occur due to the interaction between machines and the
pressure systems in which they operate. The following examples serve
to illustrate this and the fact that any change in operating conditions
should be a cause for concern.

9.2 An air compressor failure

Air compressors are widely used to supply air for utility services and
as feed stock for air separation plant. Efficiency is of prime concern.
During the 1970s a new design concept was made available. This was
the introduction of the so-called 'four-poster' that is so common
nowadays. The design consists of using a gear wheel to drive two
pinion shafts. At each end of each pinion shaft is mounted a single
impeller, so providing four stages of compression. This allows each
shaft to have a different running speed thus providing more flexibility
in the impeller selection for best efficiency. In order to approach
isothermal compression intercoolers are provided between each stage.

9.2.1 Gear failure

At one site, during the first year of operation, rumbling noises were
heard from the gearbox and the compressor was shut down.

- Inspection revealed broken and damaged teeth.
- Tooth marking showed that the gears had not maintained the
  correct contact across the tooth width.
- One of the pinion shaft bearings was found to have excessive
  clearance. This showed that the pinion shaft was not running
  parallel with the gear wheel.
- The misalignment of the gear and pinion shafts matched the tooth
  contact marking on the gear teeth.

Diagnosis of the failure

It was concluded that the gear failure was due to some tooth profile defect. The defect resulted in overloading of one of the pinion bearings. Overloading caused excessive bearing wear and the resulting misalignment was the final cause of the failure.

An intense investigation was carried out on the design and manufacture of the gears. No fault in the design was found and it was concluded that there was a possible error in manufacture, perhaps some material defect or mishap in the cutting of the teeth.

The actual cause of failure

After much fruitless investigation some lateral thinking had to be applied.

Atmospheric air contains water vapour. At some point, depending on the humidity of the air, condensate will form in the intercoolers; usually in increasing amounts towards the end of the compression process. It so happened that the plant was located near a coal-burning power station. In the 1970s no one was worried about emissions and pollution. The atmospheric air was acidic. The condensate formed in the intercoolers was acidic and severe corrosion of the intercoolers took place. In the course of time the air in the later stages of compression entrained water and debris. The entrained water and debris eroded the following impeller stage and this also resulted in the build up of deposits on the impeller. Due to uneven erosion and due to some deposits falling off during starting and stopping, the impeller became unbalanced and the centrifugal forces sustained by the adjacent bearing caused excessive wear. The excessive wear caused the misalignment. The misalignment caused the gear to fail.

This is a good example of the interdependence of pressure systems, vessels, machines, and the process. Present-day machines pay much attention to corrosion protection of intercoolers, pipework, and the provision of efficient water separation. Attention is also given to condensate hold-up and discharge facilities. The use of vibration monitoring gives early warning of impeller imbalance.

9.3 The failure of a liquid oxygen pump to pump

Oil refineries in countries abroad were often built on sites remote from other industries. Cylinders needed to be filled with the oxygen gas required for cutting and welding; this oxygen had to be produced on site. To reach the required gas filling pressure liquid oxygen has to be pumped to that pressure before it is vaporized and heated to a gas, which is then filled into steel cylinders at near-atmospheric temperature.

9.3.1 Failure to pump

A liquid oxygen generator had a reciprocating pump mounted underneath that was supposed to pump liquid. The pump, driven by an electric motor and gearbox, would reciprocate but it was found that it was not delivering any liquid at the delivery point.

Failure investigation

The oxygen generating plant had been supplied as a package unit, complete with reciprocating pump. The pump had been tested and reported to perform satisfactorily on the west coast of the USA. The unit was required for service in the Middle East, where ambient temperatures can be very high. As the pump was inspected and found to be in good order it was concluded that the malfunction could only be caused by the difference in ambient temperature.

The available net positive suction head at the face of the piston in the pump would depend on how much heat could flow from the outside. The body of the pump was very well lagged and no ice was visible when liquid oxygen flowed through the pump to get it as cold as possible before starting the pump in motion.

The actual cause of failure

After reviewing the cross-section drawings of the pump and its thermal insulation it was concluded that the only significant path for heat to travel from the outside atmosphere to the face of the piston was along the piston rod. The heat would tend to raise the temperature of the liquid oxygen at the face of the piston, so during the piston withdrawal stroke more liquid could not be sucked into the pump. A heat break was therefore designed, made and fitted to the piston rod, and the pump performed satisfactorily throughout the year for many years thereafter.

9.4   A problem of the twenty-first century

There is major difference in electricity consumption between night and day. When this can no longer be accommodated by economical load reduction then two-shifting of some power stations will be required. Two-shifting is the need for stopping and starting twice a day to meet peak demand. The original combined cycle gas turbine (CCGT) plants were designed down to a minimum cost for base load and at first they escaped the need for two-shift operation. Due to the changing nature of fuel supplies and the effects of deregulation, CCGT plants are now called upon to operate on two shifts.

Two-shift operation when imposed caused reliability problems with the heat recovery steam generator (HRSG). Tube failures and fractures of headers became a frequent occurrence. This shows how a change in operating conditions will cause unexpected failures that were not considered in the original design. The problems that ensued also provide a good example of risk-based inspection. But first some investigation into the effects of two-shifting had to be undertaken to understand the nature of the problem and to find solutions.

9.4.1 Two-shift operations
CCGT plant consists of two main power generation stages. In the first stage a gas turbine, with associated generator, generates electricity. The exhaust gas from the gas turbine is used to produce steam in a HRSG. The steam is then used to drive a steam turbine connected to a further generator. There is a basic incompatibility between the gas turbine (GT), the HRSG, and the steam turbine (ST). This is because of the difference in the thermal mass of these three items. On start-up from cold, the HRSG will take longer to get up to operating temperature than the GT; it will then generate steam at a faster rate than that required by the ST for start-up. Conversely on shutdown, the HRSG will take longer to cool than the other machines. On base-load operation, stops and starts may only take place a few times a year. Two-shift operation could increase this to over 700 times a year.

Figure 9.1 shows the arrangement of a two-stage vertical HRSG. The hot exhaust gas (about 600 °C) from the GT enters at the bottom of the boiler via an exhaust duct and exits from the top. There is a high-pressure (HP) section and a low-pressure (LP) section complete with economizers to gain the maximum possible heat recovery. It should be noted that the first tube bank is that of the superheater where the highest temperature of the steam cycle is needed. At part-load and during start-up exhaust gas temperatures will approach well over 650 °C.

9.4.2 The investigation [reference (1)]
Investigations have been carried out in the USA and in the UK on the HRSG failures that have occurred. The first requirement was to

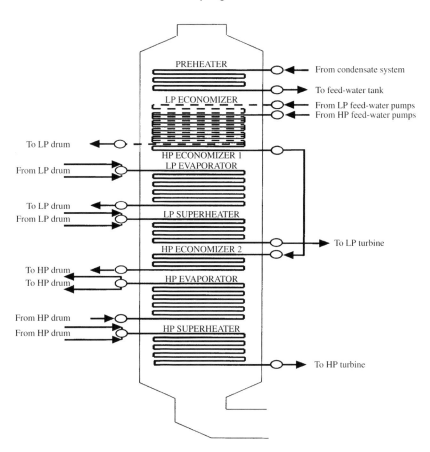

Fig. 9.1   The arrangement of a typical two-stage vertical HRSG

understand the operating environment that two-shifting imposes on the HRSG. After fitting some hundreds of thermocouples, temperature plots were obtained from various locations in the HRSG during two-shift operation and all the other operating modes. Based on these plots some selected points are shown in Table 9.1 The plots showed many peaks and troughs and some step changes in temperature. The points shown in the table have been selected to show the rapid changes in temperature that were being experienced in the boiler. Of even greater significance are the changes in steam temperature. It will be noted that the drop in steam temperature seen in the superheater will result in condensation of the steam. Any condensate will lay on the bottom of the horizontal tubes so that there will be a difference in temperature between the top length of tube compared with the

Table 9.1   Spot temperatures during two-shift operation

				Temperature ($^\circ$C)	
Time	Event	Gas	Steam	Superheater outlet T-piece	Superheater outlet header
1.44	On load	600	500	490	500
2.00	Trip	600	500	490	500
2.06			370	460	460
2.15	1st purge		310	410	410
3.15	2nd purge		250	390	320
8.00	Stabilized	100	90	100	90
9.00	Purge	100	90	100	90
9.20	Ignite	100	90	100	90
9.31			220	100	110
9.42	Synchronized		360	160	180
10.15	Steam flow		420	300	360
10.37	On load	600	500	490	500

bottom due to the difference in the mass being heated. This is further accentuated due to the fact that the hot gas will first impinge on the bottom side of the tubes.

As a result of these investigations one important modification was to add draining facilities for the superheater and evaporator headers. Any water laying in the bottom half of the headers will increase the differential expansion between the top and bottom and so add to the stresses involved. These thermal impact effects are very damaging, especially in areas of high stress concentration. Many detail modifications have to be carried out in the repair of damaged components so as to reduce stress concentration where possible. During repairs the opportunity will also need to be taken to add facilities for internal examination.

During start-up steam needs to be by-passed due to the low demand from the ST during its starting cycle. The need to use water to control the excess steam temperatures is a further hazard to be faced by the HRSG.

9.4.3   Investigation results

As a result of the investigation it has been possible to explain why two-shift operation results in HRSG component failures and to identify the failure mechanisms that are the cause. This is a good example of risk-based inspection.

The expected failure mechanisms have been identified as:

- Creep.
- Fatigue, low cycle fatigue (LCF).
- Creep fatigue.
- Corrosion:
  - oxidation due to the presence of oxygen that has not been removed;
  - dew point corrosion due to corrosive components in the flue gas being condensed;
  - erosion-corrosion (flow accelerated corrosion; FAC);
  - stress corrosion cracking (SCC);
  - corrosion fatigue.

The inspection methods to find them have been selected:

- Video probes to enter headers and tubes.
- Magnetic particle (MP) examination.
- Ultrasonic (US) examination.
- Ultrasonic thickness measurement (USTM).
- Replication – obtaining a negative by the application of a plastic film onto a surface prepared for microstructure examination. Enables non-destructive metallurgical examination of boiler components.

The inspection requirements show which failure mechanisms might occur, where they might occur and how to find them. Table 9.2 shows a general inspection plan and Table 9.3 shows one that is specific to the superheater. Figure 9.2 shows the arrangement of the HP superheater.

9.5 Summary

The above examples serve to underline the fact that the safety and reliability of machines and equipment cannot be considered in isolation. They each impact on the other in different ways and require careful consideration during design.

The law requires that equipment must be designed to be safe for all foreseeable conditions of operation. This and the requirement that all designs must be reliable and safe are dependent on the technology and experience available when a particular design is being conceived. It may also be limited by the imagination and foresight of the engineer or the owner, as in the case of the HRSG.

Table 9.2 General inspection summary

Component	Damage mechanisms	Visual	MPI	USTM	UT	Replication
HP superheater	Creep, LCF – thick section headers, also tube/header welds, possible SCC if any carry over of water	Internal video, probe general external conditions. Tubeplates and supports	*	Heaters, return bends	*	Any defects found by NDT
HP evaporator	LCF, corrosion problems if water treatment upsets occur	*	*	*		*
HP drum	Thermal fatigue, pitting	Circumferential, longitudinal, and nozzle welds, chemical injection nozzle, general internal condition	*	Shell plates and heads		*
HP economizer	Thermal fatigue, internal corrosion if incorrect water treatment, possible problems with structural rigidity	*	*	*		*
LP superheater	Caustic cracking/SCC if carry over occurs	*	*	*		*
LP evaporator	Only find problems if water treatment is of concern	*	*	*		*
LP drum	Thermal fatigue, pitting	As HP drum	*	*		*
LP economizer	FAC, thermal fatigue, pitting, dew point corrosion	*	*	*		*
Condensate preheater	Dew point corrosion, thermal fatigue, pitting if oxygen is not fully removed/not enough time to vent, FAC	*	*	*		*

*The inspection is required.

Table 9.3   HP Superheater inspection requirements

Activity	Work scope and remarks
Visual	Internal video probe of headers
	General external condition
	Tubeplates and supports
MPI	18 inlet and 18 outlet stub welds
UT	
USTM	Thickness checks on headers and return bends
Replication and hardness	Any defect locations revealed by visual testing and NDT
Damage mechanisms	Creep, LCF, thick section headers, also tube/header welds, possible SCC if any carry over

Pre-inspection activity requirements: suitable access and staging to be provided; preparatory grinding to be carried out before NDT; hand holes to be opened.

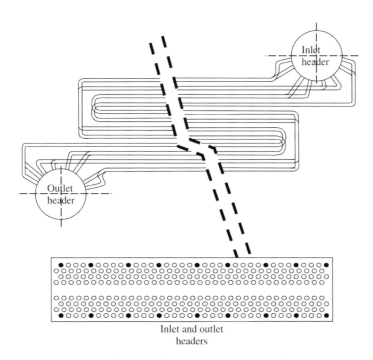

Inlet and outlet
headers

Fig. 9.2   HP superheater arrangement diagram

The need to learn as things evolve will therefore be forever present. Any new development is a challenging and high-risk business.

9.6 References

(1) **Lant, T.** and **Dometakis, C.** (2003) ERA Technology Ltd *Cyclically Operated HRSGs – Examples of Observed Damage Mechanisms.* Presented at the HRSG User Group Forum, Birmingham 2003.

Thanks are due to ERA Technology Ltd for the supply of diagrams and data, and for their permission to use them in this chapter.

Bibliography

Andrews, J. D. and **Moss, T. R.** (2002) *Reliability and Risk Assessment*. Professional Engineering Publishing, ISBN 1 86058 290 7.

Bloch, H. P. and **Geitner, F. K.** (1990) *Machinery Reliability Assessment*. Van Nostrand Reinhold, New York, ISBN 0 442 23279 9.

Canning, J. and **Ridley, J.** (1998) *Safety at Work*. Butterworth-Heinemann, Oxford, ISBN 0 7506 4018 9.

Carter, A. D. S. (1986) *Mechanical Reliability*, second edition. Macmillan, London, ISBN 0 333 40587 0.

Carter, A. D. S. (1997) *Mechanical Reliability and Design*. Macmillan, London, ISBN 0 333 69465 1.

Davidson, J. and **Hunsley, C.** (Eds) (1994) *The Reliability of Mechanical Systems*, second edition. Professional Engineering Publishing, ISBN 0 85298 881 8.

The Engineering and Technology Board. *Guidelines on Risk Issues*.

Health and Safety Executive, UK. *Free Guides*, as listed in their books catalogue.

IMechE Seminars 2000–1. *Reliability of Sealing Systems for Rotating Machinery*. Professional Engineering Publishing, ISBN 1 86058 245 1.

Kletz, T. A. (1992) *Hazop and Hazan*. IChemEng, ISBN 0 85295 285 6.

Kletz, T. A. (2001) *What Went Wrong? Case Studies of Process Plant Disasters*. Butterworth-Heinemann, Oxford, ISBN 0 8841 5920 5.

Kletz, T. A. (1991) *An Engineer's View of Human Error*, second edition. IChemEng, ISBN 0 85295 265 1.

Lees, F. P. (1996) *Loss Prevention in the Process Industries*. Butterworth-Heinemann, Oxford, ISBN 0 7506 1547 8.

Moubray, J. (1991) *Reliability-Centred Maintenance*. Butterworth-Heinemann, Oxford, ISBN 0 7506 0230 9.

Occupational Safety and Health Administration, USA. *Guides to OSHA standards*, available on their website.

O'Connor, P. D. T. (1995) *Practical Reliability Engineering*, third edition (Revised). John Wiley, Chichester, ISBN 0 471 96025 X.
Wong, W. (2002) *How Did That Happen? Engineering Safety and Reliability*. Professional Engineering Publishing, ISBN 1 86058 359 8.
Vance, J. M. (1988) *Rotordynamics of Turbomachinery*. John Wiley, ISBN 0 471 80258 1.

Department of Trade and Industry

Guidance notes on UK regulations
Electrical Equipment URN 95/626.
Electromagnetic Compatibility.
Equipment and Protective Systems Intended for use in Potentially Explosive Atmospheres URN 02/609.
Hazardous Fluids.
Machinery URN 95/650.
Pressure Equipment URN 99/1147.
Recommendations on the Transport of Dangerous Goods.

The full text of UK legislation for the above regulations and others can be obtained from H.M. Stationery Office (for address see Directory below).

HSE publications (a selection)

L122 *Safety of Pressure Systems, Approved Code of Practice.*
L22 *The Provision and Use of Work Equipment Regulations, Approved Code of Practice and Guidance.*
INDG218 *A Guide to Risk Assessment Requirements.*
HSG142 *Dealing with Offshore Emergencies.*

There are many more titles listed in the HSG books catalogue.

Informative BSI/ISO standards

BS 4778-32: 1991 *Quality Vocabulary.*
BS 5760-0: 1986 *Reliability of Systems, Equipment and Components.* An introductory guide to reliability.
BS EN 13463-1 *Non-electrical Equipment for Potentially Explosive Atmospheres.*
BSI PDI 5304 *Safe Use of Machinery.*

BS IEC 61882:2001 *Hazard and Operability Studies (HAZOP Studies), Application Guide.*
ISO IEC 60300 *Dependability Management. Part 1 Management Systems. Part 2 Guidance for Dependability Management.*
ISO IEC 60812 *Analysis Techniques for System Reliability, Procedure for Failure Mode and Effects Analysis (FMEA).*
ISO IEC 61025 *Analysis Techniques for System Reliability, Fault Tree Analysis (FTA).*
ISO IEC 61078 *Analysis Techniques for Dependability, Reliability Block Diagram Methods.*
ISO IEC 61160 *Formal Design Review.*

API standards (a selection)

Std 610 *Centrifugal Pumps for Refinery Services.*
Std 611 *General-Purpose Steam Turbines for Refinery Service.*
Std 612 *Special-Purpose Steam Turbines for Refinery Service.*
Std 616 *Combustion Gas Turbines for General Refinery Service.*
Std 617 *Centrifugal Compressors for General Refinery Services.*
Std 618 *Reciprocating Compressors for Petroleum, Chemical and Gas Industry Services.*
Std 619 *Rotary Type Positive Displacement Compressors for Refinery Service.*

Videos

(Produced by the Mechanical Reliability Committee, IMechE, and obtainable – cost on application – from the Continuing Education Unit, University of Manchester School of Engineering, The University, Oxford Road, Manchester M13 9PL, UK.)
Dissecting System Failures. Fault-tree analysis, technique, and computerized application.
Exploring Failure Consequences. Failure mode and effects analysis, technique, and applications.
Learning from failures. Weibull reliability analysis, technique, and applications.

Directory

Government agencies

European Commission, 8 Storeys Gate, London, SW1. *For information on EEC Council Directives on Health and Safety* see websites www.cec.org.uk and http://europa.eu.int/comm/employment_social/h&s/index_en.htm

Health and Safety Executive, Rose Court, 2 Southwark Bridge, London, SE1 9HS. *For information on UK Laws and Regulations on Health and Safety* see www.hse.gov.uk. and www.hsedirect.com

H.M. Stationery Office, 119 Kingsway, London, WC2. *The full text for all UK legislation, including those on Health and Safety, can be found on their website* www.hmso.gov.uk

Occupational Safety and Health Administration, USA. *For addresses of local offices and the download of OSHA Guides to Standards* see www.osha.gov

US Coast Guard. *Reports on the Human Element in Safety.* See reports and studies and the Human Element Bibliography Resource Page, www.uscg.mil/humanelements

Professional bodies

American Institute of Chemical Engineers. www.aiche.org

American Petroleum Institute. www.api.org

American Society of Mechanical Engineers. www.asme.org

American Society of Safety Engineers. www.ASSE.org

European Safety Reliability and Data Association, Beeckzanglaan lc, 1942 LS Beverwijk, The Netherlands. www.vtt.fi/aut/tau/network/esreda/esr_home.htm

Institute of Petroleum. www.petroleum.co.uk

Institution of Chemical Engineers, Davis Building, 165–171 Railway Terrace, Rugby, Warwickshire, CV21 3HQ. http://icheme.chemeng.ed.ac.uk

Institution of Mechanical Engineers. www.imeche.org.uk

National Fire Protection Association (USA). www.nfpa.org

Safety and Reliability Society, Clayton House, 59 Piccadilly, Manchester, M1 2AQ. www.sars.u-net.com

Safety Engineering and Risk Analysis Division (SERAD), ASME International. www.asme.org/divisions/serad

National organizations

British Safety Council, 70 Chancellors Road, London, W6 9RS.

The Engineering and Technology Board, 10 Maltravers Street, London, WC2. www.engc.org.uk

Fire Protection Association UK, Bastille Court, 2 Paris Gardens, London, SEI 8ND. www.thefpa.co.uk

The Loss Prevention Council, Building Research Establishment, Bucknells Lane, Garston, Watford, WD25 9XX. www.bre.co.uk

Royal Society for the Prevention of Accidents, Edgbaston Park, 353 Bristol Road, Edgbaston, Birmingham, B5 7ST. www.rospa.co.uk

Notified bodies, certifying and inspection agencies

BASEEFA (EECS). *UK certifying authority for electrical and mechanical equipment and protective systems used in flammable atmospheres, and for other electrical safety related certifying requirements. Also known as EECS, has mutual recognition agreement with Factory Mutual, USA.* www.baseefa.com

Bureau Veritas. *Notified body and other technical services.* www.Bureauveritas.com

CAN Ltd, Harness Road, Altens Industrial Estate, Altens, Aberdeen. *Inspection and NTD specialist services.* www.cangroup.net

Det Norsk Veritas (DNV), Palace House, Cathedral Street, London, SE1 9DE. *Notified body and other technical services.* www.dnv.com

Lloyd's Register. *Notified body and other technical services.* www.lr.org

Royal and SunAlliance, 17 York Street, Manchester, M2 3RS. *Notified body, engineering and certification services.* www.royalsun.co.uk

TUV. *EU notified body for testing and certification of equipment to comply with EU directives. Has worldwide offices and has a memorandum of understanding with Underwriters Laboratories, USA.* TUV technical services, www.tuvglobal.com and TUV product services, www.tuv.com

United Kingdom Accreditation Service, UKAS, 21–47 High Street, Feltham, Middlesex, TW13 4UN. *Accreditation of notified bodies, certification, testing, and calibration services.* www.ukas.com

Consultants/engineering services

ABB Eurotech Process Solutions, Daresbury Park Business Centre, Daresbury, Warrington, Cheshire, WA4 4BT. www.abb.com/service

Advantage Technical Consulting, The Barbican, East Street, Farnham, Surrey, GU9 7TB. www.advantage-business.co.uk

AEA Technology, Risley, Warrington, Cheshire, WA3 6AT. www.aeat.co.uk

AMEY Vectra, 310 Europa Boulevard, Gemini Business Park, Westbrook, Warrington, Cheshire, WA5 5YQ. www.ameyvectra.co.uk

ALSTOM. *Power station, management, operations, and maintenance services.* www.power.alstom.com

BMT Reliability Consultants Ltd, Fernside, 12 Little Farm Road, Farnham. www.bmtrcl.com

DNV Technica Ltd, Palace House, 3 Cathedral Street, London, SE1 9DE. www.dnv.com

Envirocare Technical Consultancy. *For all health and safety matters.* http://home.clara.net/envirocare

ERA Technology Ltd, *Asset management solutions, plant integrity, reliability, EMC, and many other services.* www.era.co.uk

Mitsui Babcock, Porterfield Road, Renfrew, PA4 8DJ. *Boiler technology, engineering, operations and maintenance services.*

PMSC Ltd. *RAMS and safety consultants and application software.* www.pmsafety.co.uk

Primatech. *Consultants and services for all OSHA requirements.* www.primatech.com

Quest Consultants Inc. *Consultants and services for all OSHA requirements.* www.questconsult.com

R M Consultants Ltd. *Risk management consultants and application software.* www.rmcnorth.demon.co.uk

RWE Innogy. *A complete range of services for the inspection, maintenance, management, and operation of power stations.* www.rweinnogy.com

Education and training

Institute of Risk Research, Vienna, Austria. www.irf.univie.ac.at

Primatech. *Consultants and services for all OSHA requirements.* www.primatech.com

Quest Consultants Inc. *Consultants and services for all OSHA requirements.* www.questconsult.com

School of Industrial and Manufacturing Science, Cranfield University, Cranfield, Bedford, MK43 0AL. *Safety, risk and reliability modules.* www.cranfield.ac.uk

SERCO Assurance, Thomson House, Risley, Warrington, Cheshire, WA3 6AT. *Safety and risk management training.* www.sercoassurance.com

University of Loughborough. *Courses on safety and reliability.* Loughborough, Leicestershire, LE11 3TU.

University of Manchester. *Maintenance/reliability engineering IGDS (Integrated Graduate Development Scheme) courses.* School of Engineering, Division of Mechanical Engineering, The Simon

Building, University of Manchester, Oxford Road, Manchester, M13 9PL.

Fire protection engineering services and equipment suppliers

Air Sense Technology Ltd. www.airsense.co.uk

AMEC. www.amec.co.uk

Chubb Fire Systems, Security House, Fiveways Business Centre, Aspen, Feltham, Middlesex, TW13 7AQ. www.chubb.co.uk

Electro-Detectors. www.electrodetectors.co.uk

EMS Group. www.emsgroup.co.uk

Hi Fog, Marioff Corporation Oy, PO Box 25, FIN 01511 Vantaa, Finland. *Watermist fire protection systems.* www.hi-fog.com

Tyco/Fire & Security (Wormald Fire Systems), Wormald Park, Grimshaw Lane, Newton Heath, Manchester, M40 2WL. www.wormald.co.uk

Tyco International. *Fire protection and fire detection equipment and services.* www.tycoint.com

Instrumentation

ADT Fire and Security, Security House, The Summit, Hanworth Road, Sunbury on Thames.

Bently Nevada (UK) Ltd, 2 Kelvin Close, Science Park, Birchwood, Warrington, Cheshire, WA3 7BL. *Machinery condition monitoring.* www.bently.com

Bruel and Kjaer. *Sound and vibration monitoring.* www.bksv.com

Entek IRD Mechanalysis (UK) Ltd, Bumpers Lane, Sealand Industrial Estate, Chester, CH1 4LT. *Machinery condition monitoring.* www.entekird.com

Honeywell SA, Bourgetlaan 3, 1140 Brussels, Belgium. *Plant monitoring and control.* www.iac.honeywell.com

Itronics. *UV, IR, and UV–IR optical flame detectors.* Via Applegate directory, www.applegate.co.uk

Malin Instruments. *Diesel engine monitoring.* www.malin.co.uk

Zellweger Analytics Ltd. *Gas and vapour detection.* www.zellweger-analytics.co.uk. Or via Applegate directory, www.applegate.co.uk

Machinery components

Compressor Products International. *Compressor piston and rod sealing, compressor valves, rods and engineering services.* www.compressor-products.com

Flowserve. *Pumps, control valves and mechanical seals.* www.flowserve.com

John Crane International. *Engineered sealing systems.* www.johncrane.com

Renold Hi Tec Couplings, 112 Parkinson Lane, Halifax, HX1 3QH. *Manufacture and supply of all types of couplings. Torsional vibration analysis technical services.* www.renold.com

Machinery vendors

Dresser-Rand (UK) Ltd, CI Tower, St George's Square, High Street, New Malden, Surrey, KT3 4DN. www.dresser-rand.com

Flowserve Pumps, Harley House, 94 Hare Lane, Claygate, Esher, Surrey, KT10 0RB.

GE Power Systems, General Electric International Inc., The Arena, Downshireway, Bracknell, Berkshire, RG12 1PU. www.gepower.com

GE Power Systems, Oil and Gas, Pignone Engineering Ltd, The Arena, Downshireway, Bracknell, Berkshire, RG12 1PU. www.gepower.com/geoilandgas

HMD Seal/less Pumps Ltd, Brampton Road, Hampden Park Industrial Estate, Eastbourne, East Sussex, BN22 9AN. www.hmdkontro.com

Howden Process Compressors, Old Govan Road, Renfrew, PA4 8XJ. www.process.howden.com

MAN Limited, 4 Grosvenor Place, London, SW1X 7DG. www.manturbo.com

Peterbrotherhood, Werrington Parkway, Peterborough, Cambridgeshire. *Steam turbines and gas compressors.*

Siemens, Demag, Delaval Turbo Machinery, Great West House, Great West Road, Brentford, Middlesex, TW8 9DF.

Siemens Power Generation, Siemens House, Oldbury, Bracknell, Berkshire, RG12 8FZ.

Weir Pumps, 149 Newlands Road, Cathcart, Glasgow, G44 4EX.

Noise control

Acoustic Associates. www.acousticassociates.co.uk

Acoustics Profile. www.noice.co.uk

Applied Acoustics Design. www.aad.co.uk

Envirocare Technical Consultancy. *For all health and safety matters, including noise.* http://home.clara.net/envirocare

ISVR Consultancy Services. www.isvr.co.uk

Safety system suppliers

Castell Safety International Ltd, Kingsbury Road, London, NW9 8UR. *Integrated safety solutions, interlocks.* www.castell.com

Giro Engineering Ltd, Talisman, Duncan Road, Parkgate, Southampton, Hampshire, SO31 7GA. *Sheathed diesel fuel injection pipes.* www.giro.dial.pipex.com

Loadtec Engineered Systems Ltd, The Stables, Coach House, Hythe Road, Smeeth, Kent, TN25 6SP. *Safety systems for bulk liquid transfer.* www.loadtec.co.uk

Tyco Valves and Controls. *Pressure relief, control and ESD valves.* www.tycovalves.com

ELFAB Limited, Alder Road, West Chirton Industrial Estate, North Shields, Tyne and Wear, NE29 8SD. *Bursting disc engineered products.* www.elfab.com

ICS Triplex, Hall Road, Maldon, Essex, CM9 4AL. *Safety related control systems.* www.icstriplex.com

Pilz Automation Technology, Willow House, Medicot Close, Oakley Hay Business Park, Corby, Northamptonshire, NN18 9NF. *Programmable safety control systems.* www.pilz.com

Software – safety – maintenance

Advantage Technical Consulting, The Barbican, East Street, Farnham, Surrey, GU9 7TB. *Risk modelling simulation software.* www.advantage-business.co.uk

BMT Reliability Consultants Ltd, Fernside, 12 Little Farm Road, Farnham, Surrey. *Risk analysis software.* www.bmtrcl.com

Dyadem International Ltd. *A complete range of software.* www.dyadem.com

Isograph Direct. *Risk analysis software.* www.isographdirect.com

Item Software (UK) Ltd. *A complete range of software.* www.itemuk.com

Primatech. *Process hazard analysis (HAZOP, etc.) software.* www.primatech.com

Program42 Ltd/IES Products. Asset management/maintenance system software and consultancy. www.program42-ies.com in the UK and www.iespds.com in The Netherlands.

Quest Consultants Inc. *Incident-modelling software.* www.questconsult.com

Safety and Reliability Society. *Their website provides information on software and links to providers. See 'Search the site'.* www.sars.u-net.com

Software – RBI

Det Norsk Veritas (DNV), Palace House, Cathedral Street, London, SE1 9DE. www.dnv.com

LMP Technical Services, Rockleigh Court, Rock Road, Finedon, Northamptonshire, NE9 5EL. www.lmp.co.uk

Tischuk Enterprises (UK) Ltd, 52 Regent Quay, Aberdeen, AB1 2AQ. www.energyweb.net

The Welding Institute, TWI Ltd, Granta Park, Great Abington, Cambridge, CB1 6AL. www.twi.co.uk

Data sources

EIReDA, European Industry Data Handbook, Editions SFER, Paris.

OREDA (1992). *Offshore Reliability Data Handbook*. Det Norske Veritas Industri, Norge AS, DNV Technica.

RELDAT[TM] from AEA Consultancy Services, Risley, Warrington, Cheshire.

Appendix A

Glossary

ALARP As low as reasonably practicable

API American Petroleum Institute

Availability The ability of an item (under combined aspects of its reliability, maintainability, and maintenance support) to perform its required function at a stated instant of time or over a stated period of time

ATEX Atmosphere explosive

BASEEFA British Approvals Service (for equipment in explosive atmospheres)

BSI British Standards Institute

CEN European Committee for Normalization (standards)

CENELEC European Committee for Electrotechnical Standardization

Critical machine A machine that would cause an unacceptable situation should the machine or its protection system fail

Conformity assessment A review of the Machinery Directive requirements in order to identify the actions needed to conform with the regulations

DIN German National Standards (from the German)

EC European Community

EEA European Economic Area

EEC European Economic Community

EECS Electrical Equipment Certification Service (by BASEEFA)

EEMUA Engineering Equipment and Materials Users Association

EH&S   Environment, health and safety

ESD   Emergency shutdown

Essential Health and Safety Requirements (EHSRs)   This is a list of requirements such as instructions, controls, adequate lighting, and so on; applicable to a whole range of possible hazards such as mechanical, electrical, and noise hazards

FMEA   Failure mode and effects analysis

FTA   Fault-tree analysis

HAZAN   Hazard analysis

HAZOP   Hazards and operability study

HSE   Health and Safety Executive

IEC   International Electrotechnical Commission

Inspection   The process of measuring, examining, testing, gauging, or otherwise comparing the unit with the applicable requirements

IP   Institute of Petroleum

ISO   International Standards Organization

Machine classification   In a given plant or process it is normal practice to subdivide the plant into a number of units or sub-units as convenient. The conformity assessment can be carried out on all of them so that any machinery identified can be classified in order to prioritize the hazard potential of each and to identify their critical components

Maintenance   The combination of all technical and associated administrative actions intended to retain an item in, or restore it to, a state in which it can perform its required function

NACE   National Association of Corrosion Engineers

NDT   Non-destructive testing

NEC   National Electrical Code (USA)

NEMA   National Electrical Manufacturers Association (USA)

NFPA   National Fire Protection Association (USA)

OSHA   Occupational Safety and Health Administration (USA)

PEEK   Polyetheretherketone

PFD   Process flow diagram (otherwise known as a 'flow sheet')

PTFE   Polytetrafluoroethylene

P&ID   Piping and instrumentation diagram (otherwise known as an engineering line diagram)

QA   Quality assurance

QC   Quality control

RAMS   Reliability, availability, maintenance, and safety

Registration   The process of machine classification, identification, verification, recording, and inspection to maintain the viability of the process

Reliability   The probability that a machine will perform its prescribed duty without failure for a given time when operated correctly in a specified environment

Reliability-centred maintenance (RCM)   A maintenance strategy drawn up via a structured framework of analysis aimed at ensuring the attainment of a system's inherent reliability. This results in a criticality ranking of maintenance operations based on component contribution to the overall reliability

Safety integration   The elimination of the identified risks by good design instead of 'add on' guards and safety devices

Type test   The verification, by testing and approval by a competent or notifiable authority, that an inactive unit meets all safety requirements

Work equipment   Work equipment covers any machinery, appliance, apparatus, or tool and any assembly of components, which, in order to achieve a common end, are arranged and controlled to function as a whole

Appendix B

Materials safety data sheet – some
common hazards

MATERIAL SAFETY DATA SHEET	PREP BY			
	APPD BY			
	DATE			
	ISSUE	1	2	3

SECTION 1– MATERIAL IDENTIFICATION

MANUACTURERS NAME:

CHEMICAL NAME:

TRADE NAME:

CHEMCIAL FAMILY: MOLECULAR WEIGHT:

FORMULA:

SECTION 2 – HAZARDOUS INGREDIENTS (SEE NOTE 2)

	%	TLV-TWA	TLV-STEL	TLC-CL

SECTION 3 – PHYSICAL DATA

BOILING POINT:	FREEZING POINT			
VAPOUR PRESSURE @ 20°C:	SPECIFIC GRAVITY (WATER=1)			
VAPOUR DENSITY (AIR=1):	PERCENT VOLATILES BY VOLUME:			
SOLUBILITY IN WATER:	EVAPORATION RATE:			
APPEARANCE AND ODOUR:				
CHEMICAL NAME				

	PREP BY			
	APPD BY			
MATERIAL SAFETY DATA SHEET	DATE			
	ISSUE	1	2	3

SECTION 4 – FIRE AND EXPLOSION HAZARD DATA

FLASH POINT:		
FLAMMABILITY LIMITS IN AIR (% BY VOL):	LFL=	UFL=
AUTOIGNITION TEMPERATURE:		
EXTINGUISHING MEDIA:		
SPECIAL FIRE FIGHTING PROCEDURES:		
UNUSUAL FIRE FIGHTING PROCEDURES (e.g. DUST):		
FIRE AND EXPLOSION RISK:		
MAXIMUM RATE OF PRESSURE RIST/St CLASS:		

SECTION 5 – HEALTH HAZARD DATA

THRESHOLD LIMIT VALUES:	TLV-TWA=	TLV-STEL=	TLV-CL=
SAX HAZARD RATING:			
EFFECTS OF OVER-EXPOSURE			
EMERGENCY AND FIRST AID PROCEDURES:			

SECTION 6 – REACTIVITY DATA

STABILITY:	
CONDITIONS TO AVOID:	
INCOMPATIBILITY, CORROSIVITY (MATERIALS TO AVOID):	
HAZARDOUS DECOMPOSITION PRODUCTS:	
HAZARDOUS POLYMERIZATION:	
CHEMICAL NAME:	

MATERIAL SAFETY DATA SHEET	PREP BY			
	APPD BY			
	DATE			
	ISSUE	1	2	3

SECTION 7– SPILL OR LEAK PROCEDURES

STEPS TO BE TAKEN IF MATERIAL RELEASED OR SPILLED:

WASTE DISPOSAL METHOD:

SECTION 8 – SPECIAL PROTECTION INFORMATION

RESPIRATORY PROTECTION, SPECIFY TYPE:
VENTILATION:
LOCAL EXHAUST:
SPECIAL:
MECHANICAL REQUIREMENTS (e.g. SPARK PROOF):
OTHER:
PROTECTIVE GLOVES (STATE MATERIALS):
EYE PROTECTION:
OTHER PROTECTIVE EQUIPMENT:

SECTION 9 – SPECIAL PRECAUTIONS

PRECAUTIONS TO BE TAKEN IN HANDLING AND STORING:

PRECAUTIONARY LABELLING:

OTHER PRECAUTIONS:

MATERIAL SAFETY DATA SHEET	PREP BY			
	APPD BY			
	DATE			
	ISSUE	1	2	3

CHEMICAL NAME

SECTION 10 - MISCELLANEOUS SECTION FOR ADDITIONAL INFORMATION

1. REFERENCES (ENSURE THAT LATEST EDITIONS ARE CONSULTED)

    (1A) SAX-DANGEROUS PROPERTIES OF INDUSTRIAL MATERIALS - (6TH EDITION)

    (1B) FIRE PROTECTION GUIDE ON HAZARDOUS MATERIALS - NFPA - (1985 EDITION)

2. TLV-TWA   TIME WEIGHT AVERAGE REFERENCED TO 8 HOURS EXPOSURE

   TLV-STEL SHORT-TERM EXPOSURE LIMIT REFERENCED TO 10 MINUTES EXPOSURE

   TLC-CL   CEILING CONCENTRATION SHOULD NOT BE EXCEEDED EVEN INSTANTANEOUSLY

Some common hazardous materials

Name	Formula	State	Hazard	Control	Notes
Air	$N_2\ O_2$	Pressurized gas	Death/injury	Safety valve Training/education	Rupture of container results in explosion, rupture can be from over-pressure caused by mal-operation, weakening as a result of corrosion or overheating in a fire. Detachment of a loose fitting results in a lethal missile. Jet impingement can penetrate skin into blood vessel and cause heart failure due to air bubbles in the bloodstream.
		Pneumatic conveying	Fire/explosion	Segregation from ignition sources	When used for the movement of powders, which in themselves are harmless, can become hazardous, e.g. flour and air can explode.
Oxygen	O_2	Gas	Fire	Segregation from ignition sources	Highly reactive – even metal will burn with intense heat, only needs a low energy source, such as an impact to ignite. Denser than air.
Nitrogen	N_2	Gas	Will not support life	Training/education	Commonly used to exclude air.
Carbon monoxide	CO	Colourless gas without odour	Fire/explosion	Segregation from ignition sources	LEL 12.5% UEL 74%. Auto-ignition temp. 608 °C. Danger of combustion when exposed to a flame.
			Death	Ventilation/gas detection	Asphyxiation. Due to affinity with haemoglobin low concentration causes dizziness, tiredness and headache.
Carbon dioxide	CO_2	Gas	Death	Training/education	Commonly used as a fire-fighting medium to suppress combustion. Need to evacuate in the event of release. Death caused by lack of oxygen.

Some common hazardous materials (continued)

Name	Formula	State	Hazard	Control	Notes
Halon (green equivalents)		Gas	Death or brain damage	Training/education	Commonly used as a fire-fighting medium to suppress combustion. Has the advantage of needing a lower concentration than CO_2. Will therefore support life; however, will break down into toxic products due to the heat of combustion from a fire. Breathing of toxic gas causes brain damage.
Hydrogen sulphide	H_2S	Gas	Toxic, inflammable, explosive	Gas detection	Occurs naturally in natural gas and crude oil and is also used in industrial processes. Heavier than air and so will accumulate in low lying areas. Toxic, causes irritation and respiratory paralysis. In low concentrations smells of bad eggs, exposure will kill the sense of smell; 10 p.p.m. is dangerous.
Phosgene	$COCl_2$	Gas	Highly toxic	Ventilation Breathing apparatus	Used for a wide range of industrial processes for making dyes, pharmaceuticals, etc.; 0.1 p.p.m. is dangerous. Heavier than air.
Diesel fuel		Liquid	Benign	Keep away from fire Personal hygiene	Will sustain a fire but will not easily ignite. Contact with skin can cause dermatitis.
Sulphuric acid	H_2SO_4	Liquid	Highly reactive, destroys human tissue	Containment of all spills Protective clothing Safety eye wash and showers	Widely used for industrial processes. Although non-combustible could cause a fire after coming in contact with wood or similar substances. When diluted with water and after reacting with metals will produce H_2, which can explode. Fumes are dangerous when inhaled and the liquid is dangerous to flesh and tissue.

Amelioration of hazards

Hazard	Action	Notes
Fire	Segregation	Separate sources of ignition from inflammable materials, detect inflammable gas before inflammable/explosive concentrations are reached and isolate the source of gas leak.
	Detection	Sense light emission of flames, heat radiation, rate of temperature rise.
	Escape	Provide clearly marked, unblockable, multiple means of escape.
	Control	Provide means of extinguishing and control from spreading the fire.
	Contain	Contain spread of inflammable materials and provide means of isolation such as fire-proof doors.
	Rescue and succour	Trained teams, protective clothing, breathing apparatus, means of communication, medical facilities.
Enclosed spaces, pits and cellars	Training	There is always danger of asphyxiation or fainting, fire, or explosion from accumulated gas. Such spaces must be avoided in design or provided with adequate ventilation.
	Access	Where entry is required, adequate facilities, such as opening size, platform, stairs, etc. should be provided to allow entry with all safety equipment on, and to allow rescue if required. Uncontrolled access should be prevented by locks and permit-to-work procedures.
	Testing	Atmospheres must always be tested before entry.
	Rescue	Standby help with breathing apparatus must always be at hand before entry. The person entering should have safety ropes for retrieval.
Moving parts	Separation	Safety guards and protective clothing with no items that could penetrate the guards such as long hair or loose clothing.
	Secure	Safety devices and procedures to prevent motion before removal of guards and entry.
Emergency shutdown	Containment, isolation, and release	Contain, isolate, and/or dispose of all sources of dangerous fluids. Vent all pressurized containers. Controlled stop of all operations.
Electricity/radiation, toxic gas and other hazards	Separation	Mechanical separation and shielding.
	Isolation	Work permit systems, i.e. procedures for making safe and ensuring isolation of the danger before allowing work to proceed.
	Detection	Monitoring of a safe environment both before and during work operations.

Appendix C

Machinery Directive – checklist for process machinery

Compliance checklist – Part 1

A full list with some notes and comments considered applicable for process machinery are given. The full text of everything to be considered for each of the EHSRs can be found in the DTI Machinery Guidance notes, which should be consulted when in doubt.

Notes

Machinery as used for process plant, such as pumps, compressors, turbines and engines are Group 1 machines and need only to comply with Part 1.

If they are mobile units, Part 3 EHSRs will apply. If they are intended for working underground then Part 5 EHSRs will apply.

Lifting accessories such as lifting beams for rotors, etc. must comply with the appropriate sections of Part 4 EHSRs.

Procedure

- Review the equipment design and scope against the checklist.
- If anything is not applicable, write 'N/A' in the 'APP.' (applicable) column.
- If anything is to be the responsibility of the purchaser, write 'pur.' in the APP. column.
- If anything is lacking and needs design action, write 'req.' in the APP. column and the file reference number of the action taken.
- If the purchaser is required to complete compliance then a declaration of incorporation must be issued.

1	ESSENTIAL HEALTH AND SAFETY REQUIREMENTS	APP.	File No.
1.1	**General remarks**		
1.1.1	Definitions		
	1. Danger zone means an area of risk to health and safety 2. Exposed person is anyone near or in a danger zone 3. Operator means anyone using, servicing, or moving machinery		
1.1.2	Principles of safety integration		
(a)	Machinery must be made to be safe for its intended use. The risk of accidents must be eliminated for its normal and forseeable abnormal use for its lifetime, to include assembly and dismantling.		
(b)	Risks must be avoided or reduced by design. Where it is unavoidable then protection methods must be used. Users must be warned of any residual risks and the actions needed for their avoidance.		
(c)	The machine must be designed to prevent any danger from abnormal use. The user instructions must warn of the dangers from all the possible ways of misuse.		
(d)	Ergonomic design to maintain operator efficiency must be applied.		
(e)	The design of controls must take into account any impediment due to the need for operators to wear protective equipment.		
(f)	Machinery must be supplied with any required special tools and equipment needed for operating and maintenance purposes.		
1.1.3	Materials and products		
	The use of safe materials in construction and the control of risks from fluids processed and the dangers from filling, draining, etc. *Materials of construction and sealing systems will be to API standards by the vendor.* *Process areas will be subject to DSEAR with maintenance work controlled by work permits. All by the user.*		
1.1.4	Lighting		
	The need for adequate lighting for all possible use and maintenance operations. *Applicable for machines supplied with enclosures. Possibly for local panels. Otherwise this is by the purchaser.*		

1.1.5	Design of machinery to facilitate handling		
	Machinery and each component part must have facilities for safe handling and storage. *API standards. Require lifting lugs for removable casings. Special tools are supplied for maintenance removal and lifting operations. Spare rotors have to be supplied suitable for safe storage for 3 years.*		
1.2	**Controls**		
1.2.1	Safety and reliability of controls		
	These are usually designed to API standards with variations as agreed with the buyer. Process machinery usually operates unattended with no need for operator intervention. Controls are concerned with safeguarding the machine and the process.		
1.2.2	Control devices		
	Control devices that need operator intervention are those for the starting of steam turbines; also manual unloading devices on reciprocating compressors. Gas turbines start up on automatic sequence and their control systems are proven by experience. Common to all machines is the need to start up and control utilities such as oil systems. These do not represent a risk to operators, only to the machine.		
1.2.3	Starting		
	This requires the need for a positive action to start and to avoid any risk from providing controls to start from more than one position. *Start signals are usually interlocked with permissive start signals from the process and the driven machinery. Where there is more than one location for a start signal there has to be a selection switch to arm the authorized location with an indication light.*		

1.2.4	Stopping		
	Need for a normal stop and an emergency lock-out stop. Need to ensure interlinking with all connected processes and machines. *Remote emergency stop push buttons are usually by the purchaser.*		
1.2.5	Mode selection		
	The machine must be able to operate safely for all modes of operation. If certain safety devices can be neutralized then operations must be restricted to ensure safety. *Modes of operation of process machinery are to do with changing process conditions or capacity. In some cases control panels are designed to allow instrument signals to be neutralized to allow maintenance. Such operations pose a danger to the machine and must be assessed to ensure that there is no risk.* *In other cases enclosures, such as for engines and gas turbines, are needed to control noise or risk of fire. Operator entry may require the need to disable the fire protection system and provisions are required to ensure safety.*		
1.2.6	Failure of the power supply		
	The interruption, or re-establishment, or fluctuation of the power supply must not cause the machine to malfunction. It must not be prevented from stopping after a trip signal nor must it restart of its own volition. *Control panels for critical machines are usually connected to uninterruptible power supplies.*		
1.2.7	Failure of the control circuit		
	A fault in the control circuit or failure or damage to the circuit must not lead to a dangerous situation.		

1.2.8	Software		
	The man–machine software interface must be user friendly.		
1.3	**Protection against mechanical hazards**		
1.3.1	Stability		
	There must be no risk of overturning, falling, or unexpected movement.		
1.3.2	Risk of break up during operation		
	Turbine blade failures and coupling failures are rare. Vibration monitoring is provided to alarm and trip. Piping failures on reciprocating compressors are less rare and need to be monitored.		
1.3.3	Risk due to ejected parts		
	The casing usually contains turbine blade failures.		
1.3.4	Risks due to surfaces and sharp edges or angles		
1.3.5	Risks related to combined machinery		
	Provision needed to allow safe operation of the driver without connection to the process machine.		
1.3.6	Risks relating to variations in the rotating speed of tools		
	In the case of process machinery with variable speed drives, danger due to over-speed or running on a critical speed. There could be a consequential risk to the operator. API standards specify the required safeguards.		
1.3.7	Prevention of risks due to moving parts		
	Moving parts must be safeguarded from operator intervention. If a blockage occurs then there should be a safe way for dealing with it. *Probably the only part that is exposed on process machinery will be the valve gear on a steam turbine, usually out of reach of the operator.*		

1.3.8	Choice of protection against risks related to moving parts		
	Design of coupling guards is specified in API standards. Note that a load test on guards is required by the harmonized standards.		
1.4	**Required characteristics of guards and protection devices**		
1.4.1	General requirements		
	They must be robust, at an adequate distance from the moving part, and need tools to remove.		
1.4.2	Special requirements for guards		
	• They must be securely fixed and if possible be unable to remain in place when their fixings are removed. • Movable guards must remain fixed to the machine and have safety interlocks to prevent start-up when open and to trip if opened when the machine is running. • Adjustable guards restricting access to those areas of the moving parts strictly necessary for the work. *These are required for reciprocating compressors to allow limited access to the flywheel for barring over.*		
1.5	**Protection against other hazards**		
1.5.1	Must be safe from electrical dangers		
1.5.2	Must be safe from static electricity		
	Usually earthed but in rare cases the machine may itself generate static electricity. If found, earth brushes may be needed.		
1.5.3	Energy supplies other than electricity must be safe; hydraulic, pneumatic, thermal, etc.		
1.5.4	Error of fitting		
	API standards require measures to ensure the correct mating of parts and the avoidance of erroneous assembly.		

1.5.5	Extreme temperature		
	People must be protected from these risks, also from the discharge of materials at extreme temperatures. Machines can be supplied with suitable insulation of extreme temperature surfaces but the risk of discharge will be the responsibility of the plant designer.		
1.5.6	Fire		
	Machines must be designed to avoid all possible risk of fire from overheating, etc. *The risk of fire produced by a process machine will be due to overheated bearings. Bearing temperature detectors are only optional per API.*		
1.5.7	Explosion		
	Machines must be designed to avoid any risk of explosion posed by the machine itself. *Reciprocating compressors and diesel engines have a risk of crankcase explosions. See API 618, para. 2.9.7 option. Oil mist detectors can also be used.*		
1.5.8	Noise		
	The level of noise control required depends on the length of exposure of operators and is determined by the plant designer.		
1.5.9	Vibration		
	This must be reduced to the minimum possible.		
1.5.10	Radiation		
	Any emissions must be limited so that the health of exposed persons is completely unaffected.		
1.5.11	External radiation		
	Machinery must be designed so as to be unaffected.		

1.5.12	Laser equipment		
	Any equipment used together with any optical equipment must be designed to avoid any health risk. Any accidental radiation, effective radiation, secondary radiation, or that reflected or diffused must be avoided. *Laser equipment is often used for machine alignment.*		
1.5.13	Emissions of dust, gases, etc.		
	The safe collection and disposal of vents and drains are the responsibility of the plant designer.		
1.5.14	The risk of being trapped in a machine		
	This has to be controlled by a suitable work permit system by the operator.		
1.5.15	Risk of slipping, tripping, or falling.		
	Parts of the machine where persons are liable to move about or stand must be designed and constructed so as to avoid the risk of slipping, tripping, or falling.		
1.6	**Maintenance**		
1.6.1	Machinery maintenance		
	It must be possible to carry out all maintenance and servicing with the machine stopped. Machinery maintenance areas must all be outside danger zones. *Gas turbines, for example, have enclosures around a danger zone. Operator entry may be necessary during operation. Features to ensure safety will need to be provided and agreed with the authorities. See paragraph 1.2.5.*		
1.6.2	Access to operating positions and service points		
	In the case of process plant, the manufacturer may not always provide every means of access. Where responsibility is split then this must be made clear and recorded in the declaration of incorporation.		

1.6.3	Isolation of energy sources		
	To ensure safety during maintenance operations all energy sources must be isolated. *In the case of process machinery these facilities are usually part of the process plant scope. See above.*		
1.6.4	Operator intervention		
	Any operator intervention needed should be free of risk. *Examples are lubricator adjustment on reciprocating compressors and the filling of barrier fluids for pressurized dual-seal arrangements for pumps.*		
1.6.5	Cleaning of internal parts		
	Process machinery handles dangerous substances. The plant designer provides the venting and purging of any inventory. This must be listed in the declaration of incorporation. However, in the case of reciprocating compressors the directive requires safe entry into the crankcase to be the manufacturer's design responsibility. Note that API 618, para 3.11.5 provides safety devices only as a purchaser's option.		
1.7	**Indicators**		
1.7.0	Information devices		
	These must be clear and easy to understand. Audio and light signals must give warning of any danger.		
1.7.1	Warning devices		
	These must be unambiguous and clear. Lights must be of the required colours. Facilities to allow the operator to check the operation of warning devices at any time must be provided.		

1.7.2	Warning of residual risks		
	Provide warning notices on the equipment to alert the operator to residual risks. These must be pictograms or in the language of the country of use.		
1.7.3	Marking		
	CE marking and ATEX marking as applicable. *Nameplate data are required per API standards.* Clearly mark with its mass any item that needs lifting.		
1.7.4	Instructions		
	The instruction manual must contain the following: • A repeat of the required CE nameplate information. • All information for installation, operation, maintenance, etc. *Contents generally as specified by API standards.* • Information on how the machine should be used as well as how it should not be used. • Any instructions with regard to ATEX operations if applicable. • The equipment sound pressure level; if above 85 dB(A) also the sound power level. • Machinery must also be accompanied by an instruction manual in the language of the country where it is to be used.		

Appendix D

Documentation relating to a fire-water pump

Contents

- CE-type name plate.
- CE-type data sheet.
- Technical file index.
- Compliance checklist.
- HAZID and HAZOP.
- Documentation (sample).

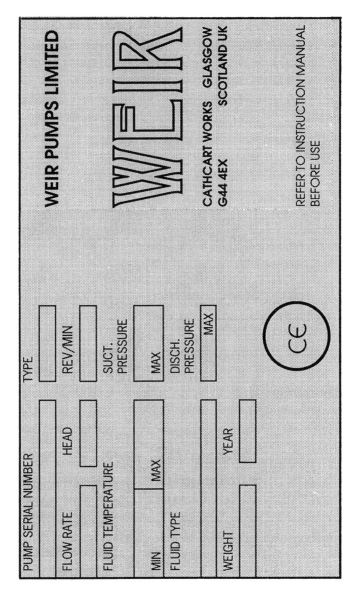

Example of CE-type name plate

WEIR PUMPS LTD   PUMP DATA SHEET

ISSUE NUMBER :   02
DATE OF ISSUE :   28-Apr-95
PREPARED BY :   S.BRADSHAW   CONTRACT SPECIFIC DATA

PUMP FRAME SIZE :	HL 50-32-200
PUMP SERIAL NUMBER :	68038-019/20
DRIVER TYPE AND DESIGNATION :	MOTOR D132S

GENERAL ARRANGEMENT DRAWING NUMBER :	A1-604230
SECTIONAL ARRANGEMENT DRAWING NUMBER :	N/A

CONTRACT DUTY HEAD :	36 m
CONTRACT DUTY FLOW :	18 m³/hr
CONTRACT DUTY SPEED :	2890 rev/min
CONTRACT IMPELLER DIAMETER :	181 mm

HYDRAULIC/MECHANICAL LIMITATIONS DATA
ALLOWABLE LIMITS FOR SAFE OPERATION

MAXIMUM PUMP OPERATING SPEED	3000 rev/min
MINIMUM NPSH REQUIRED :	2.4 m
MAXIMUM SUCTION PRESSURE :	1.0 Barg
MAXIMUM DISCHARGE PRESSURE :	4.3 Barg
MAXIMUM CONTINUOUS PUMPED FLUID FLOWRATE :	21.6 m³/hr
MINIMUM CONTINUOUS PUMPED FLUID FLOWRATE :	4.5 m³/hr
MAXIMUM PUMPED FLUID VISCOSITY :	1.519 mm²/s
MINIMUM PUMPED FLUID VISCOSITY :	0.477 mm²/s
MAXIMUM PUMPED FLUID DENSITY :	1000 kg/m³
MINIMUM PUMPED FLUID DENSITY :	983 kg/m³
MAXIMUM OPERATING TEMPERATURE (PUMPED FLUID) :	60 °C
MINIMUM OPERATING TEMPERATURE (PUMPED FLUID) :	5 °C
MAXIMUM OPERATING TEMPERATURE (AMBIENT AIR) :	25 °C
MINIMUM OPERATING TEMPERATURE (AMBIENT AIR) :	0 °C

ALLOWANCES AND LIMITATIONS ON PUMPED FLUID TYPES

FRESH WATER WITH A PARTICLE COUNT OF <150 PPM AND PARTICLE SIZE OF <75 MICRONS
NOT TO BE USED TO PUMP TOXIC FLUIDS
-
-
-
-

```
UNCONTROLLED COPY
```

ENVIRONMENTAL LIMITATIONS

NOT TO BE USED IN AN EXPLOSIVE ENVIRONMENT
-
-
-
-

DERIVED FROM TEST RESULTS ON SIMILAR EQUIPMENT

SOUND EMISSION DATA

SOUND PRESSURE LEVEL :	76 dB(A)
SOUND POWER LEVEL :	92 dB(A)

The noise is given as a maximum sound pressure level round the equipment, Lp, re 2 × 10^{-5} N/m² at one metre from the equipment running on its own in the free-field over a reflecting half plane and as a sound power level of the equipment, Lw, re 10^{-12} Watts

Example of a CE pump data sheet

PRODUCT TECHNICAL FILE

Product designation: SBWM 690 fire pumpset
Contract number: 13168
Issue number: 1

CONTENTS

1)   Safety declaration certificate

2)   The supply of machinery (safety) regulations
        1992 schedule 3 compliance checklist

Appendix 1	List of standards used in the pump design
Appendix 2	Pump design checklist
Appendix 3	Pump test certificates: – Hydrotest certificate – Rotor balance certificate
Appendix 4	Pump data sheet
Appendix 5	Pump instruction manual
Appendix 6	Pump sectional arrangement drawing Pump GA drawing Pump parts list Pump component parts drawings P&I drawing Logic diagram
Appendix 7	Systematic hazard identification study (HAZID)
Appendix 8	The identification of hazards in operation (HAZOP)

**THE SUPPLY OF MACHINERY
(SAFETY) REGULATIONS 1992
SCHEDULE 3
ESSENTIAL HEALTH AND SAFETY REQUIREMENT RELATING TO THE DESIGN
AND CONSTRUCTION OF MACHINERY**

COMPLIANCE CHECKLIST

TOPIC	APP	COMMENTS	FILE REF
1.1 GENERAL REMARKS			
1.1.2 PRINCIPLES OF SAFETY INTEGRATION	YES	Complies HAZIDS and HAZOPS used to identify all possible hazards and safety risks.	Appendices 1-8
1.1.3 MATERIALS AND PRODUCTS	YES	Materials used are checked for toxicity and safety in use. Provision for draining is provided where required at engine and gearbox.	Appendix 4 Pump data sheet Appendix 6 Sectional arrangement drawing Parts list GA Drg
1.1.4 LIGHTING	NO	Not required.	
1.1.5 DESIGN OF MACHINERY TO FACILITATE ITS HANDLING	YES	* GA Drg to indicate unit lifting points. Instruction manual to detail lifting procedure. Erection drawing details how to lift pump.	Appendix 5 Instruction manual Appendix 6 Casing Drg Impeller Drg Shaft Drg GA Drg Erection Drg
1.2 CONTROLS			
1.2.1 SAFETY AND RELIABILITY OF CONTROL SYSTEMS	YES	* Control panel supplier to confirm compliance with 1.2.1-1.2.8.	Appendix 6 Logic Description P&I Drg
1.2.2 CONTROL DEVICES	YES	See 1.2.1	
1.2.3 STARTING	YES	See 1.2.1	
1.2.4 STOPPING DEVICE	YES	See 1.2.1	
1.2.5 MODE SELECTION	YES	See 1.2.1	
1.2.6 FAILURE OF POWER SUPPLY	YES	See 1.2.1	

TOPIC	APP	COMMENTS	FILE REF
1.2.7 FAILURE OF THE CONTROL CIRCUIT	YES	See 1.2.1	
1.2.8 SOFTWARE	YES	See 1.2.1	
1.3 PROTECTION AGAINST MECHANICAL HAZARDS			
1.3.1 STABILITY	YES	Consideration of stability has been included during design. Holding down points for baseplate are provided. Gearbox and engine bolted to baseplate. Pump is supported inside casson.	Appendix 6 GA Drg
1.3.2 RISK OF BREAKUP DURING OPERATION	YES	Design calculations shown that pumpset design is sound if operated within stated limits.	Appendix 2 Pump design checklist Appendix 4 Pump data sheet
1.3.3 RISKS DUE TO FALLING OR EJECTED OBJECTS	YES	Pump and motor casings will contain any fragments.	Appendix 6 Pump casing drawing Motor casing Drg
1.3.4 RISKS DUE TO SURFACES, EDGES, OR ANGLES	YES	Drawings of components with exposed surfaces show no risk.	Appendix 6 Section arrangement drawing GA Drg Pump casing drawing
1.3.5 RISKS RELATED TO COMBINED MACHINERY	NO	Pump has only one defined operating mode.	
1.3.6 RISKS RELATED TO VARIATIONS IN ROTATIONAL SPEED OF TOOLS	NO	Speed controlled by engine governor. Overspeed trip provided on engine.	Appendix 5 Instruction manual
1.3.7 PREVENTION OF RISKS RELATED TO MOVING PARTS	YES	Wire mesh guard fitted over exposed shaft at coupling and gland, coupling guard fitted between engine and gearbox.	Appendix 6 GA drawing Parts list

TOPIC	APP	COMMENTS	FILE REF
1.3.8 CHOICE OF PROTECTION AGAINST RISKS RELATED TO MOVING PARTS	YES	Pumpset is fitted with a fixed guard at the gland and coupling locations in compliance with Section 1.4.2.1.	Appendix 6 GA drawing Parts list
1.4 REQUIRED CHARACTERISTICS OF GUARDS AND PROTECTION DEVICES			
1.4.1 GENERAL REQUIREMENT	YES	From experience the guards fitted at the pump gland and coupling locations comply with this section.	Appendix 6 GA drawing Parts list
1.4.2 SPECIAL REQUIREMENTS FOR GUARDS			
1.4.2.1 FIXED GUARDS	YES	Guard fitted at the gland and coupling locations with screws.	Appendix 6 Parts list
1.4.2.2 MOVABLE GUARDS	NO	No movable guards are fitted.	
1.4.2.3 ADJUSTABLE GUARDS RESTRICTING ACCESS	NO	No adjustable guards are fitted.	
1.4.3 SPECIAL REQUIREMENTS FOR PROTECTION DEVICES	NO	None fitted.	
1.5 PROTECTION AGAINST OTHER HAZARDS			
1.5.1 ELECTRICAL SUPPLY	YES	* Control panel supplier to confirm design code used. Other voltages present are not hazardous.	Appendix 2 Pump design checklist Appendix 6 P&I Drg
1.5.2 STATIC ELECTRICITY	YES	All components of the pumpset are earthed to the baseplate.	Appendix 6 GA Drg and Parts list
1.5.3 ENERGY SUPPLY OTHER THAN ELECTRICITY	YES	* Engine supplier to confirm design code used.	

TOPIC	APP	COMMENTS	FILE REF
1.5.4 ERRORS OF FITTING	YES	From experience there is minimal risk.	Appendix 6 GA drawing Instruction manual
1.5.5 EXTREME TEMPERATURES	NO	Pumpset is supplied without lagging as agreed in the contract. Customer must lag all exposed areas with a surface temperature in accordance with EN 563.	
1.5.6 FIRE	NO	Pump is not considered a fire risk. * Engine supplier to confirm what measures are taken to prevent fire.	
1.5.7 EXPLOSION	NO	Pumpset is not suitable for use in an explosive environment.	
1.5.8 NOISE	YES	Reasonable consideration taken.	Appendix 4 Pump data sheet
1.5.9 VIBRATION	YES	Reasonable consideration taken. Impeller balance.	Appendix 3 Pump impeller balance certificate
1.5.10 RADIATION	NO	Pumpset operation does not involve any possible emission of radiation.	
1.5.11 EXTERNAL RADIATION	NO	Pumpset will be designed to be unaffected by external radiation if specified by the customer at time of order. Otherwise standard environmental conditions apply.	Appendix 4 Pump data sheet
1.5.12 LASER EQUIPMENT	NO	No laser equipment is incorporated in the pump.	
1.5.13 EMISSION OF DUST, GASES, ETC.	YES	Any leakage from pump seal can be drained using drain connection provided.	Appendix 4 Pump data sheet Appendix 6 GA drawing
1.6 MAINTENANCE			
1.6.1 MACHINERY MAINTENANCE	YES	Complies except for gland adjustment operation. * Instruction manual to detail safe adjustment of the gland while pump is running.	Appendix 5 Instruction manual Appendix 6 GA Drg

TOPIC	APP	COMMENTS	FILE REF
1.6.2 ACCESS TO OPERATING POSITION AND SERVICING POINTS	NO	Additional access not required.	Appendix 6 GA Drg
1.6.3 ISOLATION OF ENERGY SOURCES	YES	* Customer to provide means of isolating electrical supply. * Provision for isolation of main battery supply to be considered.	Appendix 6 P&I Drg
1.6.4 OPERATOR INTERVENTION	YES	Reduced as much as possible within the scope of pump design.	Appendix 6 GA drawing
1.6.5 CLEANING OF INTERNAL PARTS	YES	Internal parts can be cleaned by circulation of a cleaning medium through the pump.	Appendix 6 Section arrangement drawing Pump casing drawing
1.7 INDICATORS			
1.7.0 INFORMATION DEVICES	YES	* Control panel supplier to confirm compliance.	Appendix 6 Logic description P&I Drg
1.7.1 WARNING DEVICES	YES	* Control panel supplier to confirm compliance.	Appendix 6 Logic description P&I Drg
1.7.2 WARNING OF RESIDUAL RISKS	YES	* Warning of gland adjustment required in instructions and on machinery.	Appendix 5 Instruction manual
1.7.3 MARKING	YES	* Name plate to be reviewed to ensure all safety information has been included e.g. maximum speed suction pressure and temperature.	
1.7.4 INSTRUCTIONS	YES	* Instructions to be checked for compliance.	Appendix 5 Instruction manual

HAZARD IDENTIFICATION SHEET DATA

Study title: HAZID STUDY

System/Area: FIREPUMP

Date: 1.6.94

Participants
G FOY - CONTRACTS
G DONALDSON - C&I
P E McFADDEN - DO
F C PORTEOUS - WES
A GREIG - S&S

A FERGUSON - TEST
L MAXWELL - DESIGN

Sheet 1 of 6

Ref No.	Generic hazard and guideword	Event description	Consequence/escalation	Prevention/control mitigating factor	Risk index Cons frequency	Action required/comment
1	Gas ingress	Gas in safe area	Gas in safe area	Leak-tight seal/welds and cable glands	Low	Gas-tight seals and welds (BRV/WPL) Cable glands (WPL)
2	Gas ingress	Room contaminated with exhaust fumes Danger to personnel	Room contaminated with exhaust fumes. Danger to personnel	Leak-tight seals on exhaust pipework. Partially tested during string test	Low	Gas-tight seals on exhaust pipework (BRV)
3	Gas ingress	Room contaminated with explosive gas	Room contaminated with explosive gas	Insufficient quantity to produce danger	Nil	No action required. Handled by normal HVAC system
4	Gas ingress	Leakage from air start supply cylinders	No dangerous emissions. Low pressure in air start cylinders. Air start inoperative	Low pressure alarm. Switch set to alarm with sufficient air remaining in cylinders to start the engine	Medium	Check system during routine maintenance (BRV)
5	Gas ingress	Gas in diesel engine air intake	Engine overspeed. High exhaust gas temperature. Engine damage	Overspeed protection	Low	Air intake ducting to be routed from safe area (BRV)

HAZARD IDENTIFICATION SHEET DATA

Study title: HAZID STUDY

System/Area: FIREPUMP

Date: 1.6.94

Participants
G FOY - CONTRACTS
G DONALDSON - C&I
P E McFADDEN - DO
F C PORTEOUS - WES
A GREIG - S&S

A FERGUSON - TEST
L MAXWELL - DESIGN

Sheet 2 of 6

Ref No.	Generic hazard and guideword	Event description	Consequence/ escalation	Prevention/control mitigating factor	Risk index Cons frequency	Action required/ comment
6	Gas ingress	Exhaust gas emissions setting off fire alarm systems	Emergency procedures instigated. Loss of production	Exhaust pipe - work routed to an area away from alarms	Medium	Gas-tight seals and welds (BRV/WPL) Cable glands (WPL)
7	Gas ingress	Gas leakage into enclosure	Pockets of gas under skid. Possible explosion	Integrity of floor penetrations	Low	Gas-tight seals on exhaust pipework (BRV)
8	Fire (jet)	Fuel line rupture	Jet fire	Heavy schedule fuel lines plus remotely operated fuel shut-off valve	Medium	No action required. Handled by normal HVAC system
9	Fire (pool)	Leakage from fuel tank	Potential pool fire	Drain pan on tank sited under potential leakage source piped away via baseplate drain pan to drainage area	Low	Check system during routine maintenance (BRV)
10	Fire (flash)	Failure of temperature control switch on gearbox lube oil heater	Flash fire	Maximum temperature attainable is calculated at 65% of oil flash point. Heater would fail prior to the flash temperature being reached	Low	Choice of switch with higher cut off or a high-temperature switch with alarm available. WPL happy with existing arrangement. BRV to action change if required

HAZARD IDENTIFICATION SHEET DATA

Study title: HAZID STUDY

System/Area: FIREPUMP

Date: 1.6.94

Participants
G FOY - CONTRACTS
G DONALDSON - C&I
P E McFADDEN - DO
F C PORTEOUS - WES
A GREIG - S&S

A FERGUSON - TEST
L MAXWELL - DESIGN

Sheet 3 of 6

Ref No.	Generic hazard and guideword	Event description	Consequence/ escalation	Prevention/control mitigating factor	Risk index Cons frequency	Action required/ comment
11	Explosion (fire)	Crankcase emissions	Ignition of gas	Crankcase of breather vented to safe area	Low	BRV to vent through roof of enclosure and to consider spark arrestor. WPL to show TP point (N) on general arrangement
12	Explosion (missiles)	Air start cylinders over-pressurized	Cylinders bursting creating missiles	Warning name plate on air start equipment giving maximum charge pressure	Medium	O & M manual to include maximum pressure within air start text (WPL)
13	Pollution	Failure of anti-fouling unit	Pump and piping fouled up or strainer blocked by marine growth	Visual indication of current failure	Low	No action required
14	Pollution	Engine air intake filter dirty/blocked	Loss of engine	Visual indicator on each filter with 'red line' warning	Low	No action required
15	Temperature	Room overheating after shutdown (instruments and cables, batteries recommended max. temp. 50 °C)	Eventual breakdown of cables and shortened battery life	Reduce temperature	Medium	WPL to estimate room temp. and effects on equipment. BRV to adequately rate the HVAC system

HAZARD IDENTIFICATION SHEET DATA

Study title: HAZID STUDY   B P ANDREW

System/Area: FIREPUMP

Date: 1.6.94

Participants
G FOY - CONTRACTS
G DONALDSON - C&I
P E McFADDEN - DO
F C PORTEOUS - WES
A GREIG - S&S

A FERGUSON - TEST
L MAXWELL - DESIGN

Sheet 4 of 6

Ref No.	Generic hazard and guideword	Event description	Consequence/ escalation	Prevention/ control mitigating factor	Risk index Cons frequency	Action required/ comment
16	Temperature	Hot equipment during running and after shutdown	Injury to operating personnel	Crankcase of breather vented to safe area	Low	WPL to provide name plates
17	Maintenance	Removing air start cylinders for recharging without closing isolating valves	Release of high-pressure air. Potential injury to personnel	Warning name plate on air start equipment giving maximum charge pressure	Low	WPL to provide name plates
18	Leakage	Joints on CW harness leakage	Loss of cooling. Failure of engine	Visual indication of current failure	Low	No action required
19	Leakage	Leakage of fuel oil	Danger to personnel due to unsound surface. Failure of unit to operate	Visual indicator on each filter with 'red line' warning	Low	No action required
20	Leakage	Battery case damaged	Batteries enclosed within battery box	Reduce temperature	Low	Check battery condition during routine maintenance (BRV/BP)

HAZARD IDENTIFICATION SHEET DATA

Study title: HAZID STUDY

System/Area: FIREPUMP

Date: 1.6.94

Participants
G FOY - CONTRACTS
G DONALDSON - C&I
P E McFADDEN - DO
F C PORTEOUS - WES
A GREIG - S&S

Sheet 5 of 6
A FERGUSON - TEST
L MAXWELL - DESIGN

Ref No.	Generic hazard and guideword	Event description	Consequence/escalation	Prevention/control mitigating factor	Risk index Cons frequency	Action required/comment
21	Location	Bel caisson and jacket caisson misalignment. Poor caisson support	Difficulty in installing pump. Movement of caisson during operation. Poor pump performance and reliability	Lead in bellmouth on caisson	Medium	WPL advise misalignment tolerances on caisson. BRV to advise caisson movements including $2g$ acceleration forces
22	Location	Sparks in exhaust gas	Ignition of external gas	Spark arrestor in silencer	Low	No action required
23	Location	Fuel shut-off. Valve cable route produces exceptionally long cable	Cable stretches when used. Shut-off valve fails to operate	Shorter cable route	Medium	WPL to advise maximum cable length
24	Location	Emergency escape from control panel region. Control panel door	Partial blockage of escape route	Location of fuel lines relative to control panel. Heavy schedule fuel lines	Low	Removal of 90° stop on control panel door
25	Location	Flames entering fuel tank via vent pipe	Escalation of fire. Fire pump inoperative	Deluge system. Fitting of flame arrestor into vent pipe	Low	BRV to consider the need for fitting flame arrestor
26	Control mechanisms	Air start button masked from view	Operator unable to locate start button. Unit fails to start	Location detailed in engine manual	Medium	WPL to provide directional name plate near to location, clearly sited

HAZARD IDENTIFICATION SHEET DATA

Study title: HAZID STUDY

System/Area: FIREPUMP

Date: 1.6.94

Participants
G FOY – CONTRACTS
G DONALDSON – C&I
P E McFADDEN – DO
F C PORTEOUS – WES
A GREIG – S&S

Sheet 6 of 6
A FERGUSON – TEST
L MAXWELL – DESIGN

Ref No.	Generic hazard and guideword	Event description	Consequence/ escalation	Prevention/ control mitigating factor	Risk index Cons frequency	Action required/ comment
27	Human factor	Manual operation selected	No automatic start	Setting indicated on control panel and main control	Low	No action required
28	Human factor	Isolation valves not opened after re-charging cylinders	No air flow to start motor. Unit fails to start	Low-pressure switch alarm	Medium	Review air start system and remove any unnecessary isolation valves
29	Human factor	Contact with rotating components	Injury to operator	Guard on all exposed rotating parts	Low	No action required
30	Human factor	Remote fuel shut-off valve closed outwith emergency condition. Not reset	No fuel to engine. Unit fails to start or operates for short time	Failed to start signal from control panel. Second unit start initiated	Medium	BRV to locate remote fuel shut-off valve mechanism within a lockable cabinet with breakable glass panel
31	Human factor	Equipment lifted using incorrect methods	Damage to equipment. Potential injury to personnel	Installation and erection drawings. O & M manual gives itemized weight schedule	Low	No action required

WEIR PUMPS LIMITED

SAFETY ACTION SHEET – MAIN HAZID

| CONTRACT NO: | 13168 |
| DESCRIPTION: | B.P. ANDREW |

| AREA OR SYSTEM: | FIRE-WATER PUMP |

SAFETY ACTION (Refer H.I.D.S.)

REF NO: H.I.D.S. REF NO. 15

BRIEF DESCRIPTION: Refer to attached sheet and appropriate 'Ref No.' for action
details.

INITIATOR: L. MAXWELL   SIGNED: _L(Maxwell_   DATE: 10/6/94

ACTION IMPLEMENTED (TO BE COMPLETED BY RECIPIENT)

FAX TO BRU/BP DATED 29/8/94 GIVING
DETAILS OF ROOM TEMPERATURES AND
INDICATING ANY EFFECTS WILL BE NEGLIGIBLE.

L MAXWELL
NAME/POSITION: DESIGN ENG SIGNED: _Maxwell_   DATE: 29/8/94

ACTION COMPLETE (TO BE COMPLETED BY INITIATOR)

INITIATOR: _LMaxwell_   DATE: 29/8/94

NOTE: On completion of 'Action Implemented Section' sheet to be returned to initiator

WEIR PUMPS LIMITED

SAFETY ACTION SHEET – MAIN HAZID

CONTRACT NO:	13168
DESCRIPTION:	B.P. ANDREW

AREA OR SYSTEM:	FIRE-WATER PUMP

SAFETY ACTION (Refer H.I.D.S.)

REF NO: H.I.D.S. REF NO. 16

BRIEF DESCRIPTION: Refer to attached sheet and appropriate 'Ref No.' for action
details.

INITIATOR: L. MAXWELL   SIGNED: *L/Maxwell*   DATE: 10/6/94

ACTION IMPLEMENTED (TO BE COMPLETED BY RECIPIENT)

INSTRUCTION ISSUED TO DRAWING OFFICE TO
PROVIDE WARNING NAMEPLATES ON BOTH
SIDES OF SKID TO HIGHLIGHT "HIGH
TEMPERATURE".

L MAXWELL
NAME/POSITION: DESIGN ENG   SIGNED: *L/Maxwell*   DATE: 13/6/94

ACTION COMPLETE (TO BE COMPLETED BY INITIATOR)

INSTRUCTIONS ISSUE BY D.O. FOR MANUFACTURE.

INITIATOR: *L/Maxwell*   DATE: 27/6/94

NOTE: On completion of 'Action Implemented Section' sheet to be returned to initiator

WEIR PUMPS LIMITED

SAFETY ACTION SHEET – MAIN HAZID

CONTRACT NO:	13168
DESCRIPTION:	B.P. ANDREW

AREA OR SYSTEM:	FIRE-WATER PUMP

SAFETY ACTION (Refer H.I.D.S.)

REF NO: H.I.D.S. REF NO. 17

BRIEF DESCRIPTION: Refer to attached sheet and appropriate 'Ref No.' for action
details.

INITIATOR: L. MAXWELL   SIGNED: _L(Maxwell_   DATE: 10/6/94

ACTION IMPLEMENTED (TO BE COMPLETED BY RECIPIENT)

INSTRUCTIONS ISSUED TO DRAWING OFFICE TO

PROVIDE NAME PLATE ON AIR START EQUIP^T TO

INSTRUCT USER TO VENT CYLINDER AND CLOSE

ISOLATION VALUE PRIOR TO REMOVAL BEFORE RECHARGING.

L MAXWELL

NAME/POSITION: DESIGN ENG .SIGNED: L(Maxwell DATE: 13/6/94

ACTION COMPLETE (TO BE COMPLETED BY INITIATOR)

INSTRUCTIONS ISSUED BY D.O. FOR MANUFACTURE

INITIATOR: L(Maxwell   DATE: 27/6/94

NOTE: On completion of 'Action Implemented Section' sheet to be returned to initiator

WEIR PUMPS LIMITED

SAFETY ACTION SHEET – MAIN HAZID

CONTRACT NO:	13168
DESCRIPTION:	B.P. ANDREW

AREA OR SYSTEM:	FIRE-WATER PUMP

SAFETY ACTION (Refer H.I.D.S.)

REF NO: H.I.D.S. REF NO. 20

BRIEF DESCRIPTION: Refer to attached sheet and appropriate 'Ref No.' for action
details.

INITIATOR: L. MAXWELL   SIGNED: _L(Maxwell_   DATE: 10/6/94

ACTION IMPLEMENTED (TO BE COMPLETED BY RECIPIENT)

FAX SENT TO BRV/BP DATED 29/8/94
HIGHLIGHTING ACTION.

L. MAXWELL
NAME/POSITION: DESIGN ENG. SIGNED: L(Maxwell DATE: 29/8/94

ACTION COMPLETE (TO BE COMPLETED BY INITIATOR)

INITIATOR: L/Maxwell   DATE: 29/8/94

NOTE: On completion of 'Action Implemented Section' sheet to be returned to initiator

WEIR PUMPS LIMITED

SAFETY ACTION SHEET – MAIN HAZID

CONTRACT NO:	13168
DESCRIPTION:	B.P. ANDREW

AREA OR SYSTEM:	FIRE-WATER PUMP

SAFETY ACTION (Refer H.I.D.S.)

REF NO: H.I.D.S. REF NO. 21

BRIEF DESCRIPTION: Refer to attached sheet and appropriate 'Ref No.' for action
details.

INITIATOR: L. MAXWELL   SIGNED: _L I Maxwell_   DATE: 10/6/94

ACTION IMPLEMENTED (TO BE COMPLETED BY RECIPIENT)

THIS ACTION HAS BEEN THE SUBJECT OF MUCH
DISCUSSION BETWEEN WPL AND BP DEV^L AND IS
DOCUMENTED ON VARIOUS FAXES. A COMPROMISED
SITUATION HAS BEEN REACHED WHICH IS ACCEPTABLE TO
BOTH. FAXES FILED IN RELEVANT JOB PACKET.

NAME/POSITION: L MAXWELL / DESIGN ENG SIGNED: _L I Maxwell_   DATE: 16/6/94

ACTION COMPLETE (TO BE COMPLETED BY INITIATOR)

INITIATOR: _L I Maxwell_   DATE: 16/6/94

NOTE: On completion of 'Action Implemented Section' sheet to be returned to initiator

WEIR PUMPS LIMITED

SAFETY ACTION SHEET – MAIN HAZID

CONTRACT NO:	13168
DESCRIPTION:	B.P. ANDREW

AREA OR SYSTEM:	FIRE-WATER PUMP

SAFETY ACTION (Refer H.I.D.S.)

REF NO: H.I.D.S. REF NO. 23

BRIEF DESCRIPTION: Refer to attached sheet and appropriate 'Ref No.' for action
details.

INITIATOR: L. MAXWELL   SIGNED: *L(Maxwell* DATE: 10/6/94

ACTION IMPLEMENTED (TO BE COMPLETED BY RECIPIENT)

FAX SENT TO BRV | BP DATED 29/8/94
HIGH LIGHTING ACTION.

L Maxwell
NAME/POSITION: DESIGN ENG SIGNED: *L(Maxwell* DATE: 30/8/94

ACTION COMPLETE (TO BE COMPLETED BY INITIATOR)

INITIATOR: *L(Maxwell*   DATE: 30/8/94

NOTE: On completion of 'Action Implemented Section' sheet to be returned to initiator

WEIR PUMPS LIMITED

SAFETY ACTION SHEET – MAIN HAZID

CONTRACT NO:	13168
DESCRIPTION:	B.P. ANDREW

AREA OR SYSTEM:	FIRE-WATER PUMP

SAFETY ACTION (Refer H.I.D.S.)

REF NO: H.I.D.S. REF NO. 24

BRIEF DESCRIPTION: Refer to attached sheet and appropriate 'Ref No.' for action
details.

INITIATOR: L. MAXWELL   SIGNED: _L / Maxwell_   DATE: 10/6/94

ACTION IMPLEMENTED (TO BE COMPLETED BY RECIPIENT)

INSTRUCTION ISSUED TO ELECTRICAL FOREMAN
TO HAVE STOP REMOVED PRIOR TO UNIT
LEAVING W PL WORKS.

L MAXWELL
NAME/POSITION: DESIGN ENG. SIGNED: _L / Maxwell_   DATE: 11/8/94

ACTION COMPLETE (TO BE COMPLETED BY INITIATOR)

INITIATOR: _L / Maxwell_   DATE: 11/8/94

NOTE: On completion of 'Action Implemented Section' sheet to be returned to initiator

FIRE-WATER PUMP

SAFETY ACTION SHEET LOG   **MAIN HAZOP**

Action No.	Date	Subject	Action
627	10 May 94	Main valve to diesel engine closed	Completed
628	10 May 94	Valve on vent left closed	Completed
629	10 May 94	Flexible connection missing between secondary air start system and the diesel engine	Completed
630	10 May 94	Block valve upstream and downstream of PVC 36004 closed	Completed
631	10 May 94	Block valve upstream and downstream of PVC 36009 closed	Completed
632	10 May 94	Block valve upstream and downstream of PVC 36008 closed	Completed
633	10 May 94	Burst tube on lube oil cooler	Pending
634	10 May 94	Consider installing PIs downstream of PCVs 36004 and 36008	Pending

Rev 0   Issued 2 June 94

HAZARD AND OPERABILITY STUDY ACTION SHEET
DATA FILE

ACTION ON: Weir - MECH	RESPOND BY:
ACTION NO: 627 MEETING DATES: 14/2/94	

DOCUMENT REFERENCE: PME06XC0501+PID-1011-001/2/D2   REVISION: 2
TITLE: FIRE-WATER PUMPS

PLANT SECTION: (HAZOP Table 631/2)
SECONDARY AIR START TO THE DIESEL ENGINE

CAUSE:
Main valve to diesel engine closed

CONSEQUENCE:
Unable to crank machine in an emergency

SAFEGUARDS/REMARKS:
None

ACTION:
Consider locking open main valve

RESPONSE: DATED: 1/6/94

THE VALVE IN QUESTION IS IN CLOSE PROXIMITY TO THE AIR START BUTTON AND IS EASILY ACCESSIBLE. IF IT WERE TO BE LEFT CLOSED THERE WOULD BE NO PROBLEM IN CORRECTING THE SITUATION AND THE TIME TAKEN NEGLIGIBLE.

A NAMEPLATE ADDED TO AIRSTART SYSTEM HIGH-LIGHTING VALVE SHOULD BE LEFT OPEN.

OPERATING AND MAINTENANCE MANUAL WILL INCLUDE TEXT WHICH HIGHLIGHTS THE IMPORTANCE OF VALVE BEING OPEN AND TO CHECK VALVE IF AIR START MOTOR DOES NOT OPERATE WHEN BUTTON IS PUSHED. THE DANGER IS A LOCKABLE VALVE MAY BE LOCKED CLOSED IN ERROR THUS ISOLATING THE AIR SUPPLY. IT IS DECIDED THEREFORE MAIN VALVE TO REMAIN UNLOCKABLE.   SIGNED: L.Maxwell

ENTER YOUR RESPONSE IN THE BOX ABOVE THEN SIGN AND RETURN THIS FORM TO:
B. Brazier

NOTES: (For use of HAZOP Secretary only)

SAFETY ACTION SHEET - MAIN HAZOP

ACTION ARISING FROM ADP-SA-REP-0552-000	ACTION NO: SAS-MHA-627

AREA OR SYSTEM: See attached sheet	Date: See attached sheet

SAFETY ACTION: (description from original report)

ACTN NO. 627   SEE ATTACHED SHEET

Signed: B'W. Brazier   Name: B.BRAZIER   Date: 4/5/94

ACTION TAKEN TO RESOLVE: PLEASE ENTER YOUR RESPONSE ON THIS SHEET

See attached

action for implementation.
check if nameplate added.

Signed: Name: Date:

This Action has been Approved:
Project Engineer: Date: 16/9/94
Lead Engineer: Date:
Safety Engineer: Date: 16/9/94

This Action has been implemented: Cost/Savings:
Safety Engineer: Date: £:

HAZARD AND OPERABILITY STUDY ACTION SHEET
DATA FILE

ACTION ON: Weir	RESPOND BY:

| ACTION NO: 628 | MEETING DATES: 14/2/94 | |

DOCUMENT REFERENCE: PME06XC0501+PID-1011-001/2/D2   REVISION: 2
TITLE: FIRE-WATER PUMPS

PLANT SECTION: (HAZOP Table 631/2)
SECONDARY AIR START TO THE DIESEL ENGINE

CAUSE:
Valve on vent left open

CONSEQUENCE:
Loss of air to diesel engine when required to start

SAFEGUARDS/REMARKS:
None

ACTION:
Consider locking vent valve closed

RESPONSE: DATED: 1/6/94
LOW PRESSURE ALARM INCORPORATED IN AIR START SYSTEM TO FLAG UP LOW PRESSURE. AIR ESCAPING FROM VENT VALVE WOULD BE AUDIBLE. NAMEPLATE ADDED TO AIRSTART SYSTEM WARNING AGAINST LEAVING VENT VALVE OPEN. IT IS CONSIDERED UNNECESSARY TO AUGMENT THE ABOVE FEATURES WITH LOCKABLE VALVES.   SIGNED: L Maxwell

ENTER YOUR RESPONSE IN THE BOX ABOVE THEN SIGN AND RETURN THIS FORM TO:
B. Brazier

NOTES: (For use of HAZOP Secretary only)

SAFETY ACTION SHEET – MAIN HAZOP

ACTION ARISING FROM ADP-SA-REP-0552-000	ACTION NO: SAS-MHA-628

AREA OR SYSTEM: See attached sheet	Date: See attached sheet

SAFETY ACTION: (description from original report)

ACTN NO. 628   SEE ATTACHED SHEET

Signed: B.W. Brazier   Name: B. BRAZIER   Date: 4/5/94

ACTION TAKEN TO RESOLVE: PLEASE ENTER YOUR RESPONSE ON THIS SHEET

See attached
action for implementation.
check if nameplate added.

Signed: Name: Date:

This Action has been Approved:
Project Engineer: ABR   Date: 16/9/94
Lead Engineer: Date:
Safety Engineer: Date: 16/9/94

This Action has been implemented: Cost/Savings:
Safety Engineer: Date: £:

HAZARD AND OPERABILITY STUDY ACTION SHEET
DATA FILE

ACTION ON: Weir ～ MECH	RESPOND BY:
ACTION NO: 629 MEETING DATES: 14/2/94	

DOCUMENT REFERENCE: PME06XC0501+PID-1011-001/2/D2   REVISION: 2
TITLE: FIRE-WATER PUMPS

PLANT SECTION: (HAZOP Table 631/2)
SECONDARY AIR START TO THE DIESEL ENGINE

ACTION:
Show flexible connection between secondary air start system and diesel engine

RESPONSE: DATED: 1/6/94

FLEXIBLE CONNECTION SHOWN ON
P & I DIAGRAM N° AO - 501045.

SIGNED: *L. Maxwell*

ENTER YOUR RESPONSE IN THE BOX ABOVE THEN SIGN AND RETURN THIS FORM TO:
B. Brazier

NOTES: (For use of HAZOP Secretary only)

SAFETY ACTION SHEET - MAIN HAZOP

ACTION ARISING FROM ADP-SA-REP-0552-000	ACTION NO: SAS-MHA-629

AREA OR SYSTEM: See attached sheet	Date: See attached sheet

SAFETY ACTION: (description from original report)

ACTN NO. 629   SEE ATTACHED SHEET

Signed: *B.W.Brazier*   Name: *B.BRAZIER*   Date: 4/5/94

ACTION TAKEN TO RESOLVE: PLEASE ENTER YOUR RESPONSE ON THIS SHEET

See attached

Signed: Name: Date:

This Action has been Approved:
Project Engineer: Date: 16/9/94
Lead Engineer: Date:
Safety Engineer: Date: 16/9/94

This Action has been implemented: Cost/Savings:
Safety Engineer: Date: 16/9/94   £:

HAZARD AND OPERABILITY STUDY ACTION SHEET
DATA FILE

ACTION ON: Weir – MECH	RESPOND BY:

ACTION NO: 630	MEETING DATES: 14/2/94

DOCUMENT REFERENCE: PME06XC0501+PID-1011-001/2/D2   REVISION: 2
TITLE: FIRE-WATER PUMPS

PLANT SECTION: (HAZOP Table 631/3)
SEA-WATER COOLING FROM PUMP DISCHARGE TO LUBE OIL COOLER, ENGINE COOLER
AND AIR COOLER

CAUSE:
Block valves upstream and downstream of PCV36004 closed

CONSEQUENCE:
High gearbox temperature and potential mechanical damage

SAFEGUARDS/REMARKS:
This is likely to be undetected

ACTION:
Consider locking open block valves

RESPONSE: DATED: 1/6/94

LOCKABLE VALVES INCORPORATED IN GEARBOX
COOLING HARNESS. LOCKED OPEN POSITION
INDICATED ON P&I DIAGRAM N° AO-50104S.
OPERATING AND MAINTENANCE MANUAL WILL
INCLUDE NECESSARY INSTRUCTIONS WITHIN
ITS TEXT.

SIGNED: L.L.Maxwell

ENTER YOUR RESPONSE IN THE BOX ABOVE THEN SIGN AND RETURN THIS FORM TO:
B. Brazier

NOTES: (For use of HAZOP Secretary only)

SAFETY ACTION SHEET – MAIN HAZOP

ACTION ARISING FROM ADP-SA-REP-0552-000	ACTION NO: SAS-MHA-630

AREA OR SYSTEM: See attached sheet	Date: See attached sheet

SAFETY ACTION: (description from original report)

ACTN NO. 630   SEE ATTACHED SHEET

Signed: B.W. Brazier   Name: B BRAZIER   Date: 4/5/94

ACTION TAKEN TO RESOLVE: PLEASE ENTER YOUR RESPONSE ON THIS SHEET

See attached

action for implementation.
Check lockable valves included
Ensure lockable valves shown on
                         P&ID

Signed: Name: Date:

This Action has been Approved:

Project Engineer: ABR   Date: 16/9/94
Lead Engineer: Date:
Safety Engineer: Date: 16/9/94

This Action has been implemented: Cost/Savings:

Safety Engineer: Date: £:

HAZARD AND OPERABILITY STUDY ACTION SHEET
DATA FILE

ACTION ON: Weir — Mech	RESPOND BY:
ACTION NO: 631	MEETING DATES: 14/2/94

DOCUMENT REFERENCE: PME06XC0501+PID-1011-001/2/D2   REVISION: 2
TITLE: FIRE-WATER PUMPS

PLANT SECTION: (HAZOP Table 631/3)
SEA-WATER COOLING FROM PUMP DISCHARGE TO LUBE OIL COOLER, ENGINE COOLER
AND AIR COOLER

CAUSE:
Block valves upstream and downstream of PCV36009 closed

CONSEQUENCE:
Increase in temperature of diesel engine and potential mechanical damage

SAFEGUARDS/REMARKS:
None

ACTION:
Consider locking open these block valves

RESPONSE: DATED: 1/6/94
LOCKABLE VALVES INCORPORATED IN COOLING
WATER HARNESS. LOCKED OPEN POSITION
INDICATED ON P&I DIAGRAM N° AO-50104S.
OPERATING AND MAINTENANCE MANUAL WILL
INCLUDE NECESSARY INSTRUCTIONS WITHIN
ITS TEXT

                                                    SIGNED: L. Maxwell
ENTER YOUR RESPONSE IN THE BOX ABOVE THEN SIGN AND RETURN THIS FORM TO:
B. Brazier

NOTES: (For use of HAZOP Secretary only)

SAFETY ACTION SHEET - MAIN HAZOP

ACTION ARISING FROM ADP-SA-REP-0552-000	ACTION NO: SAS-MHA-631

AREA OR SYSTEM: See attached sheet	Date: See attached sheet

SAFETY ACTION: (description from original report)

ACTN NO. 631   SEE ATTACHED SHEET

Signed: B.W. Brazier   Name: B. BRAZIER   Date: 4/5/94

ACTION TAKEN TO RESOLVE: PLEASE ENTER YOUR RESPONSE ON THIS SHEET

See attached

Action for implementation.
Check lockable valves included
Ensure   "   "   shown on P+ID

Signed: Name: Date:

This Action has been Approved:
Project Engineer: PBR   Date: 16/9/94
Lead Engineer: Date:
Safety Engineer: Date: 16/9/94

This Action has been implemented: Cost/Savings:
Safety Engineer: Date: £:

HAZARD AND OPERABILITY STUDY ACTION SHEET
DATA FILE

ACTION ON: Weir — MECH	RESPOND BY:

ACTION NO: 632	MEETING DATES: 14/2/94

DOCUMENT REFERENCE: PME06XCO501+PID-1011-001/2/D2   REVISION: 2
TITLE: FIRE-WATER PUMPS

PLANT SECTION: (HAZOP Table 631/3)
SECONDARY COOLING FROM PUMP DISCHARGE TO LUBE OIL COOLER, ENGINE COOLER
AND AIR COOLER

CAUSE:
Block valve downstream of PCV36008 closed

CONSEQUENCE:
No air cooling in enclosure when required during a fire

SAFEGUARDS/REMARKS:
None

ACTION:
Consider locking open this block valve

RESPONSE: DATED: 1/6/94

LOCKABLE VALVES INCORPORATED IN COOLING
WATER HARNESS. LOCKED OPEN POSITION
INDICATED ON P & I DIAGRAM N° AO-501045.
OPERATING AND MAINTENANCE MANUAL WILL
INCLUDE NECESSARY INSTRUCTIONS WITHIN ITS
TEXT.

SIGNED: L/Maxwell

ENTER YOUR RESPONSE IN THE BOX ABOVE THEN SIGN AND RETURN THIS FORM TO:
B. Brazier

NOTES: (For use of HAZOP Secretary only)

SAFETY ACTION SHEET - MAIN HAZOP

ACTION ARISING FROM ADP-SA-REP-0552-000	ACTION NO: SAS-MHA-632

AREA OR SYSTEM: See attached sheet	Date: See attached sheet

SAFETY ACTION: (description from original report)

ACTN NO. 632   SEE ATTACHED SHEET

Signed: B.W. Brazier   Name: B. BRAZIER   Date: 4/5/94

ACTION TAKEN TO RESOLVE: PLEASE ENTER YOUR RESPONSE ON THIS SHEET

See attached

action for implementation.
Check lockable valves included
Ensure   "   "   are on P+ID.

Signed: Name: Date:

This Action has been Approved:

Project Engineer: Date: 16/9/94
Lead Engineer: Date:
Safety Engineer: Date: 16/9/94

This Action has been implemented: Cost/Savings:

Safety Engineer: Date: £:

SAFETY ACTION SHEET - MAIN HAZOP

ACTION ARISING FROM ADP-SA-REP-0552-000	ACTION NO: SAS-MHA-633

AREA OR SYSTEM: See attached sheet	Date: See attached sheet

SAFETY ACTION: (description from original report)

ACTN NO. 633   SEE ATTACHED SHEET

Signed: B.W. Brazier   Name: B. BRAZIER   Date: 4/5/94

ACTION TAKEN TO RESOLVE: PLEASE ENTER YOUR RESPONSE ON THIS SHEET

Signed: Name: Date:

This Action has been Approved:
Project Engineer: Date: 16/9/94
Lead Engineer: Date:
Safety Engineer: Date: 16/9/94

This Action has been implemented: Cost/Savings:
Safety Engineer: Date: 16/9/94   £:

MECHANICAL PLANT INSPECTION REQUIREMENT – ERA TECHNOLOGY

PLANT: HRSG   COMPONENT: HP SUPERHEATER

Activity	Work scope and remarks	Man days	
		Est.	Act.
Visual	– Internal videoprobe of headers – General external condition – Tubeplates and supports		
MPI	– 18 inlet and 18 outlet stub welds		
UT			
USTM	– Thickness checks on headers and return bends		
Replication and hardness	– Any defect locations revealed by visual tests and NDT		
Damage mechanisms	Creep, LCF – thick section headers; also tube/header welds, possible SCC if carry over		

Pre-inspection activity requirements:
– Suitable access and staging to be provided   – Handholes to be opened
– Preparatory grinding to be carried out before NDT

Inlet and Outlet Headers

Index

Acceptable risk – ALARP 49, 61
Air:
  compressor failure 193
  process hazards 154
Alternating current potential drop (ACPD) 183
Anthrometrics 56
API code classifications 15
Approval of modifications 105
As low as reasonably practicable (ALARP) 49, 61
Assemblies of machines 25
Assessment:
  change of 104
  consequence 188
  modules:
    applicable 41, 46
    conformity 41
  probability 188
  risk 187
ATEX Directive 5, 18
Axial compressors 118
  characteristics of 119

Bearings 113, 121, 138
Biomechanics 56

Casings 112, 119, 138
CDM regulations 2, 4, 10
Centrifugal compressors 112
  characteristics of 117
Centrifugal pumps 137
Checks and reviews 105
Chemical contamination 151
Classification:
  diagram 91
  of machines 90
  steps 40

Client 11, 87
  as operator 12
Codes and standards 2, 7
COMAH regulations 4, 5, 49
Common:
  mode failure 71
  requirements, equipment and protective systems 18
Compliance 35
  duty of 5
  routes to 34
  with requirements 104
Compression machinery 111
Compressor:
  failure, air 193
  frame (crankcase, crosshead, and running gear) 126
  reciprocating, characteristics 136
Compressors:
  axial 118
    characteristics of 119
  centrifugal, characteristics of 117
  reciprocating 125
  screw 119
    characteristics of 123
  comparison of oil free and oil-flooded 124
Conceptual studies 76
Confidentiality 30
Conformity assessment:
  assessment:
    procedure 39
    requirements 20
  modules 41
Consequence 186
  assessment 188
Construction (Design and Management) (CDM) Regulations 1995 2, 4, 10

Construction contractor 90
Construction/commissioning engineer
  90
Containment 1, 186
  review 97
Contamination, chemical 151
Control of Major Accident Hazards
  Regulations (COMAH) 1999 4, 5,
  49
Control of Substances Hazardous to
  Health Regulations (COSHH) 1994
  3
Control system reliability 72
Controls and instruments 181
Corrosion 170
  cracking, stress 174
  galvanic 171
  microbial 173
  or erosion 184
  pitting and crevice 172
  protection 175
  type of, and their symptoms 171
Creep 168
Crossheads 129
Cryogenic pumps 142
Cylinder:
  heads, pistons, and piston rods 125
  liners 128

Dangerous area of a zone 14
Dangerous Substances and Explosive
  Atmospheres Regulations
  (DSEAR) 2002 4, 13
Defects:
  distribution of 176
  fatigue-type 183
  root causes of 177
Dependability management 108
Design:
  assumptions, verifying 140
  hazards 116, 118, 130, 152, 122
  limits 142
  lubrication 133
  non-lube 132
  safety 4
  specification 76
Designer 12, 87
Diesel engines 157

Directives 1
  ATEX 5, 18
  Electromagnetic Compatibility 5
  Equipment and Protective Systems
    Intended for Use in Potentially
    Dangerous Atmospheres
    5, 18
  Low Voltage 5
  Machinery 2, 5, 23
  Pressure Equipment 5, 36
  Simple Pressure Vessels 48
Diversity, principle of 71
Documentation 1, 31
  and marking 21
  required 17
  requirements 35
  software control 166
Drivers 144
Dry:
  gas sealing systems 114
  sealing systems 122
DSEAR 4, 13
  scope 13
Duty:
  holders 6
  of compliance 5

EH&S assessment form 105
Electrical Equipment (Safety)
  Regulations 1994 34
Electromagnetic Compatibility
  Directive 5
Enforcement of the law 6
Engineering:
  line diagram 94
  team, assignment of 165
Environmental and stress factors 68
Equipment:
  and categories 19
  group 16
  mechanical 21
Equipment and Protective Systems
  Intended for Use in Potentially
  Explosive Atmospheres
  Regulations 1996 18
Ergonomics 56
Essential Health and Safety
  Requirements (EHSR) 18, 38

Evaluation procedures 185
Examination by complex plane analysis,
    eddy current 183
Explosion 185

Factory acceptance test 77
Factory owner 3
Failure 65
    actual cause of 194, 195
    air compressor 193
    common mode 71
    diagnosis of 194
    due to temperature 168
    fluid flow induced 170
    gear 193
    investigations 195
    materials 167
    mechanisms 199
    modes 1
        choice of methods for 182
    of a liquid oxygen pump 194
    on demand 71
    piston rod 131
    probability of 68
    rate 1, 67
    risk of 66, 186
    statistics 176
    to pump 195
    vibration-induced 130
Failure modes and effects analysis
    (FMEA) 61
Fatigue 168
    thermal 168
Fatigue-type defects 183
Fault-tree analysis (FTA) 73
Feed points, typical numbers of 135
Film seals, liquid oil-lubricated 115
Fluid:
    flow induced failure 170
    properties 186
FMEA 61
    for a diesel engine 62
Fractional redundancy 72

Galvanic:
    corrosion 171
    series 172
Gas turbines 154

degradation 155
enclosure 50
power output 156
types 156
Gear failure 193
General Agreement on Tariffs and
    Trade (GATT) 9
Guided wave or long wave technology
    183

Handover 99
Harm 53
Harmonized standards 37
Hazard:
    assessment
        machine 55
        process materials 57
    rating, materials 58
    review 97
Hazard and operability (HAZOP):
    studies 62
    reviews 95
Hazardous area classification 13
Hazards 1, 53, 125
    to health and safety 153, 154, 156,
        158
    air process 154
    design 118, 122, 130, 152
    human interface 56
    identification of 54
    machine 118
    maintenance 55
    manufacturing 110
    operating 112, 118, 119, 125, 137, 148
    process 109
        from steam 145
    transient and operating location 57
    tribology 132
HAZOP studies 62
Health and safety:
    hazards to 153, 154, 156, 158
    legislation, evolution of 2
HRSG, two-stage vertical 197
Human interface hazards 56
Hydrogen embrittlement 174

Identification of hazards 54
Inlet/outlet valves 127

Inspection 103
    intervals 176
    methods 199
    pressure system 182
    on-line, methods used 184
    requirements 181
    risk-based 178
Installation 77
Instrument:
    reliability 161
    schedules 94
    systems, performance comparison of
        163
Instrumentation 115
Instruments and controls 181
Investigation 196
    procedure 183
    results 198
IP code classifications 15

Labyrinth:
    gas sealings systems 114
    seals, rotor, impeller, and balance
        drums 113
Law, enforcement of 6
Leakage 186
Legislation 1
    evolution of health and safety 2
Life-cycle cost 107
Liquid oxygen pump, failure of
    194
Long wave or guided wave technology
    183
Low Voltage Directive 5
Low-load operation 151
Lubrication 153

Machine:
    and safety components, individual
        24
    design responsibilities 89
    hazards 118
        assessment 55
    maintenance and operation 95
    manufacturer 89
    operation and protection 94
    reliability 108
    supplier's data 93

verification 92
Machinery:
    Directive 2, 5, 23
    for incorporation 26
    for process plant 32
    modifications 30
    train 144
Machines, assemblies of 25
Maintenance 1
    and operations 99 et seq
    hazards 55
    instructions 96
    provisions 164
Management of change 104
Management of Health and Safety at
    Work Regulations (MHSWR)
    1999 3
Manufacturing:
    hazards 110
    QC/QA 77
Materials:
    defects 170
    failure 167
    hazard rating 58
    of construction 46
Mitigation 16
Model safety and reliability specification
    79
Modifications 104
    approval of 105
Module requirements 47
MTBF 122, 127
    conditions 115

Non-destructive testing (NDT):
    techniques, competency in 184
    welding, and other joining processes
        48
Notified body 37
    definition of 21

Occupational Safety and Health Act,
    (OSHA) 1970 3
Off-specification operation, time limits
    for 149
On-line inspection, methods used 184
Operation 77
    part-load 142

profitable 1
Operational reliability 69

Part-load operation 142
PEEK 132
Physiology 56
Piston:
 rings 128
 rod failure 131
Pitting and crevice corrosion
 172
Planning supervisor 11
Pressure Equipment Directive
 (PED) 5, 36
Pressure systems 166
 inspection methods 182
Pressure Systems Safety Regulations
 (PSSR) 2000 4, 49
Probability 65
 assessment 188
 consequence matrix 189
 of failure, 68
 factors that affect 66
Process:
 data sheet 93
 fluids 137
 hazards 109
 from steam 145
 materials hazard assessment 57
 plant, machinery for 32
Products affected 35
Project responsibilities 88
Protective systems 16
Provision and Use of Work Equipment
 Regulations (PUWER):
 1992 22
 1998 3
PTFE 132
Pump:
 characteristics 143
 centrifugal 137
 cryogenic 142

Reciprocating compressors
 125
 characteristics 136
Redundancy 65, 70
 fractional 72

principle of 71
Regulations:
 COMAH 4, 5, 49
 Construction (Design and
  Management) (CDM) 2, 4, 10
 Control of Major Accident Hazards 4,
  5, 49
 Dangerous Substances and Explosive
  Atmospheres 4, 13
 Electrical Equipment (Safety) 34
 Equipment and Protective Systems
  Intended for Use in Potentially
  Explosive Atmospheres
  Regulations 5, 18
 Management of Health and Safety at
  Work 3
 Pressure Systems Safety 4, 49
 Provision and Use of Work
  Equipment 3, 22
 scope of 23
 Supply of Machinery Safety 5, 23
 Control of Substances Hazardous to
  Health 3
 Reporting of Injuries, Diseases and
  Dangerous Occurrences 3
Reliability 1, 65
 analysis – an illustration 81
 assessment 65 et seq, 74
 data, typical 112
 effect of test intervals on 162
 improving 116, 123, 134, 140, 152,
  153,
 in design 74
 instrument 161
 prediction 69
 specification 78
Reliability-centred maintenance (RCM)
 100
 analysis 101
Reliable operation, checklist 136
Reporting of Injuries, Diseases and
 Dangerous Occurrences
 Regulations (RIDDOR) 1995 3
Requirements, compliance with 104
Residual life assessment 184
Resistance curves, system 143
Responsibilities 87
 machine design 89

project 88
Reviews and checks 105
Ring seals, floating oil-lubricated 122
Risk 53
    assessment 14, 53, 58, 177, 187
        flow diagram 59
        level matrix 60
    management 16
    of failure 66, 186
    quantitative 67
    ranking 185
        consequence severity level 59
        likelihood 60
Risk-based inspection (RBI) 178
    management of 189
Rod packing 128
Rotors 113
    and rotor seals 120
    and wear rings 138
Routes to compliance 34

Safety 1
    file 12, 100
    in design 4
    measures 14
    or loss-prevention engineer 89
    plan 11, 99
    planning supervisor 88
Screw compressors 119
    characteristics of 123
    comparison of oil free and oil-flooded
        124
Sealing systems:
    dry gas 114
    labyrinth gas 114
Seal-less pumps, the use of 141
Seals:
    Mechanical 122
        contact 115
        oil-lubricated liquid film 115
        shaft 139
Segregation, principle of 71
Selection table, typical compressor 111
Shaft seals 139
Simple Pressure Vessels Directive
    (SPVD) 48
Site (plant) manager 90

Software:
    control documentation 166
    development 164
    management of 165
    programs 189
Spot temperatures during two-shift
    operation 198
Standards and codes 2, 7
Start-up 77
Steam:
    conditions 150
    purity:
        limits 149
        the effect of poor 148
    turbines:
        general-purpose 153
        special purpose 145
Stopping and starting 155, 122, 129, 141,
    151
Stress corrosion cracking 174
Stud bolds 129
Supply of Machinery Safety Regulations
    1992 5, 23
System resistance curves 143
Systems audits 103

Technical file 27, 100
    availability of 28
    format and contents of 28
    language of 29
    objective of 27
Tie rods 129
Time of flight diffraction (TOFD)
    182
Torsional oscillations 157
Tribology hazards 132
Two-shift operations 196

Values, inlet/outlet 127
Velocity effects 173
Verification methodology 93

Water slugs 151
Wear rings 128
Welding, NTD, and other joining
    processes 48
Written scheme of examination 180